Minerals, Rocks and Inorganic Materials

Monograph Series of Theoretical and Experimental Studies

9

Edited by

W. von Engelhardt, Tübingen · T. Hahn, Aachen
R. Roy, University Park, Pa. · P. J. Wyllie, Chicago, Ill.

Subseries:
Isotopes in Geology

Jochen Hoefs

Stable Isotope Geochemistry

With 37 Figures

Springer-Verlag New York · Heidelberg · Berlin 1973

Dr. *Jochen Hoefs*

Geochemisches Institut der Universität

D-3400 Göttingen

ISBN 0-387-06176-2 Springer-Verlag New York · Heidelberg · Berlin
ISBN 3-540-06176-2 Springer-Verlag Berlin · Heidelberg · New York

Preface

Over 10 years have passed since RANKAMA's second book, Progress in Isotope Geology, surveyed the literature on isotope abundance determinations. In the meantime the number of measurements and publications has increased enormously. Therefore, it seems necessary to summarize the knowledge in this field in the light of more recent developments.

The title of this book was chosen because the whole field of radioactive isotope geochemistry has been deliberately omitted.

The book is divided into three parts. Section A gives the theory of isotope effects and the technical background, both aspects being discussed rather briefly. The author regrets some shortcomings in the introductory section, especially in the theoretical treatment of isotope fractionation, but he has been trained mainly in earth sciences rather than in physical chemistry.

Section B gives a summary of the fractionation mechanisms affecting the most important elements — hydrogen, carbon, oxygen, and sulfur. Further, it surveys some other elements that have not yet been as thoroughly investigated.

Section C surveys the most important results from a geological standpoint. In some cases the opinions of different authors on the same subject are summarized without comment, because the field of stable isotope geology is growing so rapidly that a final answer cannot be given at the moment.

It is obvious that in writing this book, which is of the survey type, the author could not rely only on his own experiments and experience. He has therefore been careful to conserve as much of their original flavor as possible and to quote his sources.

The author believes that he has included the most significant recent contributions to the literature on stable isotopes. It was not, however, his intention to offer a compendium of all published work in this field, nor to write something like a handbook of stable isotope geochemistry.

This book is not so much for the specialist active in the field, but more for students interested in investigating it. The author's hope is that this book may also have some impact on those earth scientists who are not intensively engaged in this subject. He has tried to demonstrate the broad possibilities of stable isotope determinations in solving problems in geochemistry and cosmochemistry.

This book would not have been written without the encouragement of Prof. Dr. C. W. CORRENS. Dr. NIELSEN introduced me into the basic principles of mass spectrometry and isotope geochemistry. His advice and constructive criticism is gratefully acknowledged. Valuable discussions with Professor K. H. WEDEPOHL have supported the progress of the manuscript. Drs. H. P. SCHWARCZ, S. M. F. SHEPPARD, W. STAHL, and D. H. WELTE have kindly reviewed an early draft of the manuscript. Special thanks go to Dr. P. G. COOMER, who has improved the English considerably.

Göttingen, April 1973   JOCHEN HOEFS

Contents

A. Theoretical and Instrumental Background

I. Definition of Isotopes

Isotopes may be defined as atoms whose nuclei contain the same number of protons but a different number of neutrons. The term "isotopes" is derived from Greek (meaning equal places) and indicates that isotopes occupy the same position in the periodic table.

It is convenient to denote isotopes in the following form: $^{16}_{8}O$, where the superscript 16 represents the mass number and the subscript 8 represents the atomic number.

Isotopes can be divided into stable and unstable (radioactive) species. The number of stable isotopes is about 300; whilst over 1200 unstable ones have been discovered so far. The term "stable" is a relative one, it depends on the detection limits of radioactive decay times. In the range of atomic numbers from 0 to 83, stable nuclides of all masses except 5 and 8 are known. Only 21 elements are pure elements, in the sense that they have only one stable isotope. All other elements are mixtures of at least two isotopes. In some elements, the different isotopes may be present in substantial proportions. In copper, for example, ^{63}Cu accounts for 69% and ^{65}Cu accounts for 31%. In most cases one isotope is predominant, the others being present only in traces.

The stability of nuclides is characterized by several important rules, two of which are briefly discussed here. The first one is the so-called symmetry rule, which states that in a stable nuclide with a low atomic number, the number of protons is approximately equal to the number of neutrons, or the neutron to proton ratio, N/Z, is approximately equal to unity. In stable nuclei with more than 20 protons or neutrons, the N/Z is always greater than unity, with a maximum value of about 1.5 for the heaviest stable nuclei. The electrostatic Coulomb repulsion of the positively charged protons grows rapidly with increasing Z. To maintain stability in the nuclei, electrically neutral neutrons are incorporated more rapidly than protons (see Fig. 1).

The second rule is the so-called Oddo and Harkins rule, which states that nuclei can be classified in terms of containing even or odd numbers of protons (Z) and neutrons (N). Of the four possible combinations, the most common arrangement is even-even, the least common odd-odd, as shown in Table 1.

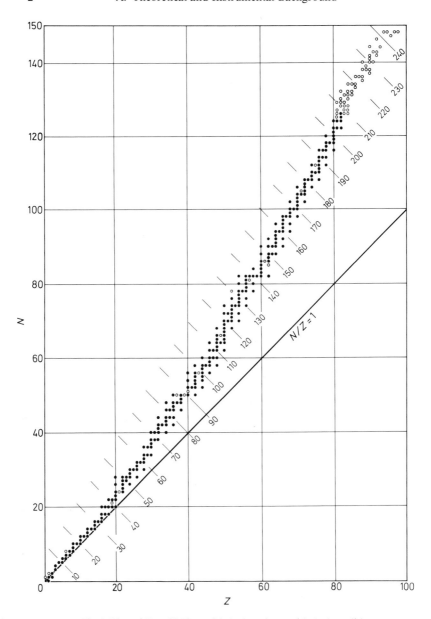

Fig. 1. Plot of Z and N in stable (●) and unstable (○) nuclides

Table 1. Types of atomic nuclei and their frequency of occurrence

Type Z/N	Number of stable nuclides
Even-even	160
Even-odd	56
Odd-even	50
Odd-odd	5

The same relationship is demonstrated in Fig. 2, which shows that there are more stable isotopes with even than with odd proton numbers.

Radioactive isotopes can be classified into artificial and natural. Only the latter are of interest in geology, because they are the basis for most age-dating methods. Radioactive processes are spontaneous nuclear reactions, characterized by the radiation emitted. This may be classified into 1) alpha radiation, 2) beta radiation, 3) gamma radiation, and 4) electron capture.

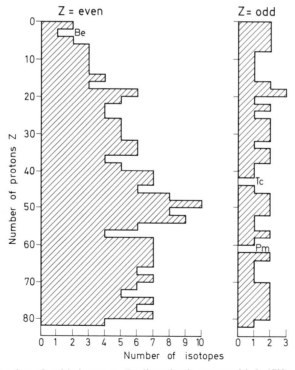

Fig. 2. Number of stable isotopes. (Radioactive isotopes with halflifes greater than 10^9 years are included)

Radioactive decay is one process that produces isotope abundance variations. The second process is that of isotopic fractionation caused by small chemical and physical differences between the isotopes of an element. It is exclusively with this process that we are dealing with in the following chapters.

II. Isotope Effects

It is well-known that the extranuclear structure of an element essentially determines its chemical behavior, whereas the nucleus is more or less responsible for its physical properties. Because all isotopes of a given element contain the same number and arrangement of electrons, a far-reaching similarity in chemical behavior is the logical consequence. But this similarity is not unlimited; there exist certain differences in physicochemical properties due to the mass differences of different isotopes. These mass differences are most pronounced amongst the lightest elements. For example, in Table 2 some differences in physico-chemical properties of H_2O and D_2O are listed.

Since the discovery of the isotopes of hydrogen by UREY et al. (1932a, b), differences in the chemical properties of the isotopes of the elements H, C, N, O, S, and other elements have been calculated by the methods of statistical mechanics and determined experimentally. These differences in the chemical properties can lead to considerable isotope effects in naturally occurring chemical reactions.

The theory of isotope effects and related isotope fractionation mechanisms will be discussed very briefly. For a more detailed introduction to the theoretical background see BIGELEISEN (1965), BRODSKY (1961), BROECKER and OVERSBY (1971), MELANDER (1960), ROGINSKY (1962), TUDGE and THODE (1950), and UREY (1947).

The quantum theory is the theoretical basis that explains the differences in the physico-chemical properties of isotopes. The energy of a

Table 2. Characteristic constants of H_2O and D_2O

Constants	H_2O	D_2O
Density (20° C, in g cm^{-3})	0.9982	1.1050
Temperature of greatest density (°C)	4.0	11.6
Mole volume (20° C, in cm^3 mole^{-1})	18.049	18.124
Melting point (760 torr, in °C)	0.00	3.82
Boiling point (760 torr, in °C)	100.00	101.42
Vapor pressure (at 100° C, in torr)	760.00	721.60
Viscosity (at 20.2° C in centipoise)	1.00	1.26
Ionic product at room temperature	1×10^{-14}	0.16×10^{-14}

molecule is described in terms of the electronic energy plus the translation, rotation, and vibration energies of the molecules. The energies associated with the mutual interactions of these motions must also be included. For isotopes of the same element the electronic, translation, and rotation energies are more or less equal. This leaves molecular vibrations as the origin of "isotope effects".

Figure 3 shows schematically the energy of a diatomic molecule as a function of the distance between two atoms. According to the quantum theory, the molecule cannot assume any energy on the continuous curve shown in Fig. 3, but is restricted to certain discrete energy levels. The

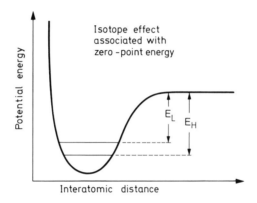

Fig. 3. Schematic potential-energy curve for the interaction of two atoms in a stable molecule or between two molecules in a liquid or solid. (After BIGELEISEN, 1965)

lowest level is not at the minimum of the energy curve, but above it by an amount of $^1/_2\, h\upsilon$, where h is Planck's constant and υ is the frequency with which the atoms in the molecule vibrate with respect to one another. The vibrational frequency of a molecule depends inversely on the masses of the atoms in the molecule. Therefore, different isotopic species will have different zero-point energies in molecules with the same chemical formula: The molecule of the heavy isotope will have a lower zero-point energy than the molecule of the light isotope. This is shown schematically in Fig. 3, where the upper horizontal line (E_L) represents the zero-point energy of the light molecule and the lower line (E_H), that of the heavy one.

This means that bonds formed by the light isotope are more readily broken than bonds involving the heavy isotope. Thus, during a chemical reaction, molecules bearing the light isotope will in general react slightly more readily than those with the heavy isotope.

Summarizing, we have seen that the energy of isotopic substances depends on the vibrational frequencies of the molecules, which in turn depends on the masses of the atoms in the molecules.

III. Isotope Fractionation Processes

The partitioning of isotopes between two substances with different isotope ratios is called isotope fractionation. The main phenomena producing isotope fractionation are:

1) isotope exchange reactions
2) kinetic processes, mainly depending on differences in reaction rates of isotopic molecules;
3) fractionations due to other physico-chemical effects.

1) Isotope exchange includes processes with very different mechanisms. In the following, the term "isotope exchange" is used for all processes in which ordinary changes in the chemical system do not occur, but in which the isotope distribution changes between different chemical substances, between different phases, or between individual molecules.

Let us consider a typical exchange reaction, an equilibrium process, which may be written as

$$aA_1 + bB_2 \rightleftharpoons aA_2 + bB_1$$

where A and B are molecules having any element as a common constituent. The subscripts 1 and 2 indicate that the molecules contain only the light or heavy molecule, respectively. Using statistical mechanics, the equilibrium constant K may be expressed in terms of the partition functions Q.

$$K = \frac{Q_{A_2}}{Q_{A_1}} \bigg/ \frac{Q_{B_2}}{Q_{B_1}}.$$

This means that the equilibrium constant K is the quotient of two partition function ratios, one for the two molecules of A, and one for B; in other words, a calculation of the partition functions allows the calculation of equilibrium constants for chemical reactions. The partition function Q of a molecule is defined by

$$Q = \sum_n g_n^{-E_n/kT}$$

where the summation extends over all the allowed energy levels E_n of the molecules and g_n is the statistical weight of the nth level E_n. BIGELEISEN and MAYER (1947) and UREY (1947) have demonstrated that, for the calculation of partition function ratios of isotopic molecules (except hydrogen, where rotational energies cannot be neglected), it is sufficient to take into account a modified vibrational partition function Q'.

The dependence of the equilibrium constant K on temperature is the most important property for geological purposes. It is interesting to note that for changes in temperature, the equilibrium constant K sometimes changes from values larger than unity to values smaller than unity and vice versa. This is known as crossing over.

Finally, approaching $0\,°K$ the equilibrium constant K tends towards zero or infinity, corresponding to complete isotope separation. This condition is never reached in practice, due to the slow rate of exchange at low temperatures. As the equilibrium differences in chemical properties of isotopes vanish at high temperatures, the effects disappear as the temperature tends towards infinity.

We have seen that, for the calculation of a partition function ratio for a pair of isotopic molecules, we have to know the vibrational frequencies of each. When solid materials are considered, the evaluation of partition function ratios becomes even more complicated, because it is necessary not only to take into account the independent vibrations of each molecule, but also to consider the lattice vibrations.

Usually we are interested in the fractionation factor, rather than in the equilibrium constant. The *fractionation factor* α is defined as the ratio of the numbers of any two isotopes in one chemical compound divided by the corresponding ratio for the other chemical species. At equilibrium, in general $\alpha = K^{1/n}$, where n is the maximum number of exchangeable atoms. But this equation does not hold true in compounds where the isotopes of hydrogen are considered, or where the different positions of several atoms of a single element are not equivalent.

Note that for a reaction such as

$$H_2{}^{18}O + 1/3\,CaC^{16}O_3 \rightleftharpoons H_2{}^{16}O + 1/3\,CaC^{18}O_3\,,$$

at equilibrium $K = \alpha$.

2) The second main phenomena producing isotope fractionations are kinetic effects. For example, in a gas phase the molecules containing the light isotope move more rapidly than those with the heavy isotope. The translational velocities of gas molecules are inversely proportional to the square root of the ratio of the molecular weights. For example, one can write for CO_2:

$$\frac{\text{velocity}\,(^{12}C^{16}O^{16}O)}{\text{velocity}\,(^{13}C^{16}O^{16}O)} = \sqrt{\frac{45}{44}} = 1.011\,.$$

This means that the velocity of CO_2 of mass 44 is 1.1% greater than that of mass 45. Such velocity differences lead to isotope separation during diffusion.

Qualitatively, many observed deviations from the simple equilibrium processes can be interpreted as consequences of the various isotopic components having different rates of reaction.

Isotope fractionation measurements taken during irreversible chemical reactions always show a preferential enrichment of the lighter isotope in the products of the reaction. The nature of this fractionation stems from the lower ground-state vibration frequency of the heavy isotope. Hence, more energy is required to destroy a molecule bearing the heavy isotope.

The transition-state theory provides the framework for any theory of kinetic isotope effects. This theory is based on the idea that a chemical reaction proceeds from some initial state to a final configuration by a continuous change and that there is some critical intermediate configuration called the activated complex or transition state.

It is further assumed that there is a small number of activated molecules in equilibrium with the reacting species and that the rate of reaction is controlled by the rate of decomposition of the activated species.

The isotope fractionation introduced in the course of a unidirectional reaction may be considered in terms of the ratio of rate constants for the isotopic substances.

For the two competing isotopic reactions

$$A_1 \xrightarrow{K_1} B_1 \quad \text{and} \quad A_2 \xrightarrow{K_2} B_2$$

the ratio of rate constants, K_1/K_2, for the reaction of light and heavy isotopic species is expressed, as in the case of equilibrium constants, in terms of two partition function ratios, one for the two isotopic reactants species, and one for the two isotopic species of the activated complex (for a more detailed discussion see MELANDER, 1960).

The third class of processes that can lead to isotope fractionation are also due to physico-chemical effects. To this category belong the following phenomena: Evaporation and condensation, crystallization and melting, absorption and desorption, diffusion and thermodiffusion. Of special interest in stable isotope geochemistry are the evaporation-condensation processes. Some of these phenomena are not necessarily different from (1) and (2) and could be also discussed there.

Differences in the vapor pressures of isotopic compounds lead to fractionations. For example, from the vapor pressure data for water it is evident that the lighter molecular species are preferentially enriched in the vapor phase, the extent depending upon the temperature. Such an isotopic separative process has been treated theoretically in terms of batch distillation of a liquid under equilibrium conditions (Rayleigh distillation). Any isotopic reaction carried out in such a way that the products are isolated immediately after formation from the reactants will show a characteristic trend in isotopic composition. Examples of this type of process are the progressive formation and removal of raindrops from a cloud, and the formation of crystals from a solution too cool to allow diffusive equilibrium between the crystal interior and the liquid. The isotopic composition of residual water vapor (or of the residual liquid) is a function of the fractionation factor between vapor and water droplets or between the liquid and surface layer of the growing solid. At any instant the vapor over the liquid is assumed to be in isotopic equilibrium with the liquid which remains, according to the relation

$$r = \alpha_0 R'.$$

The ideal fractionation factor as represented by α_0, which, for a two-component system, is equal to the ratio of the respective vapor pressures. R' refers to the molecular ratio in the liquid remaining, r to that in the vapor phase. R' changes as the distillation proceeds. A natural example of a Rayleigh distillation is the fractionation between the oxygen isotopes in the water vapor of a cloud and the raindrops released from that cloud. The resulting depletion of the $^{18}O/^{16}O$ ratio in the residual water vapor is given as a function of the fraction of original vapor remaining in the cloud (see Fig. 4).

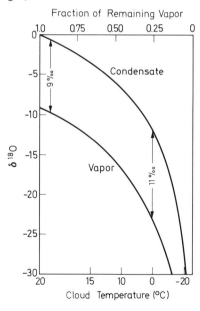

Fig. 4. $\delta^{18}O$ in cloud vapor and condensate plotted as a function of the fraction of remaining vapor in the cloud for a Rayleigh process. The temperature of the cloud is shown on the lower axis. The increase in fractionation with decreasing temperature taken into account. (After DANSGAARD, 1964). For definition of $\delta\ ^{18}O$, see p. 15.

Diffusion processes also have importance in geology on a wide scale. Diffusional separation processes for gaseous and isotopic mixtures are of two general types: Those that do not require an additional gas as a separating agent; and those in which separation of the mixture occurs by diffusion through an added gas component [carrier diffusion, CRAIG (1967)]. The first type includes the process of barrier diffusion or effusion, characterized by molecular flow through a porous barrier (e.g., the migration of methane through porous rocks) and the process of thermal diffusion across a temperature gradient.

For two isotopic species, their diffusion coefficients D and D^* are related by $D/D^* = \sqrt{M^*/M}$, where M is the mass of species. In any natural diffusive process, isotopic fractionations may arise from this relationship. The theory of isotopic fractionation by liquid or solid diffusion on geologic systems has been discussed by SENFTLE and BRACKEN (1955), who conclude that probably no large geologic body exists in which a major constituent is isotopically enriched by these diffusion processes.

In considering diffusion as a geologic process, it is important to distinguish between solid-solvent and liquid-solvent phases. The term "solid-state diffusion" generally includes the processes in which the solute diffuses through a solid matrix, along grain boundaries or over surfaces. Liquid-state diffusion refers to the diffusion of a solute through a liquid. In both types of diffusion, isotopic fractionation can occur, but to different extents. The diffusion coefficients in liquids are larger than the solid-state diffusion coefficients by several orders of magnitude. Consequently, solid-state diffusion can apply only to shortdistance phenomena. Geologic examples of isotopic fractionation caused by liquid-state or solid-state diffusion have been described (see potassium).

IV. Basic Principles of Mass Spectrometry

A number of different methods may be used for measuring the abundance of stable isotopes. The methods used are usually physical, including analysis by neutron activation, analysis of electromagnetic spectra, determination of the specific gravity of fluids, and analysis of mass spectra.

Moderately accurate determinations of the specific gravity (density) of a liquid have been described, which in favorable circumstances, such as in water, allow the determination of the isotopic composition from differences in the specific gravity of the fluids. Methods based on the characteristic electromagnetic spectra of isotopes include optical, atomic and molecular, microwave, and nuclear resonance spectrography. Methods of atomic and molecular spectroscopy are based on the existence of separable spectrum lines characteristic of the various isotopes or of isotopic molecules.

Mass-spectrometric methods are by far the most effective for measuring isotopic abundance ratios. The first mass spectrographs were built in 1918 by DEMPSTER at the University of Chicago, and in 1919 by ASTON at the Cavendish Laboratory in Cambridge, England. Recently, several types of mass spectrometers have been developed to cover the whole field of application. Much literature has been published about mass spectrometry and its application. Examples are BARNARD (1953), EWALD and HINTENBERGER (1953), INGHRAM and HAYDEN (1954),

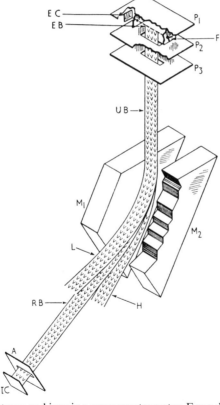

Fig. 5. Paths of electrons and ions in a mass spectrometer. For a description, see the text. Legend: *A* analyzer exit plate; *EB* electron beam; *EC* electron collector; *F* filament; *H* ions too heavy to be collected; *IC* ion collector plate; *L* ions too light to be collected; M_1, M_2 magnet poles; P_1, P_2, P_3 parallel planes; *RB* resolved ion beam; *UB* unresolved ion beam. (From RANKAMA, 1954)

RIECK (1956), DUCKWORTH (1958), McDOWELL (1963), BRUNÉE and VOSHAGE (1964), and HINTENBERGER (1966).

In principle, a mass spectrometer may be divided into four different parts: 1) the inlet system, 2) the ion source, 3) the ion deflector or mass separator, and 4) the ion detector. A diagram of the essential parts of a Nier-type 60-degree mass spectrometer is presented in Fig. 5.

A beam of electrons (EB) is emitted by a heated metal ribbon fila-ment (F), usually tungsten or rhenium, and is directed to pass between two parallel plates (P_1 and P_2). The beam is collimated by means of a weak magnetic field. Positive ions are formed as a result of collisions between the gas molecules and the electrons passing between P_1 and P_2. The ions are drawn out of the electron beam through the slit in P_2 by the action of an electric field, and are further accelerated when passing be-tween P_2 and a third plate P_3. The potential difference between P_2 and P_3 (acceleration potential) is continuously adjustable from zero to sev-eral kilovolts. The ions entering the magnetic field are essentially mono-energetic, i.e., they have a constant energy acquired in falling through the potential difference between the electron beam and the grounded plate P_3.

The positive ions with a charge e' will acquire energy equal to e' volts (eV) when passing through the electric field. After passing through the electric field, all the singly charged ions will possess the same kinetic energy:

$$1/2\,m'v^2 = e'\,\mathrm{V}\,. \tag{1}$$

The light ions will move with a high velocity, the heavy ones with a low velocity.

When the ions enter the magnetic field in a direction perpendicular to the magnetic lines of force, they will be subjected to a force perpendicular to both the direction of the field and the direction of their motion. The magnitude of this force depends on the field strength and on the charge and velocity of the ions. This force is represented by the vector equation

$$F = \frac{e'v'B}{c} \tag{2}$$

where B is the strength of the magnetic field, e' is the charge, v' is the velocity of the moving particle, and c is the velocity of light. If we combine Eqs. (1) and (2), it can be shown that the path the ion follows in the magnetic field is a function of the mass of that ion. This fact is the basis of mass spectroscopy. The ions are forced to travel in an arc, the radius of which depends on their mass and energy. Heavy ions will have greater radius values than the light ions.

The chosen ions of the resolved ion beam (RB) reconverge on and pass through a defining slit on the analyzer exit plate (A) and then fall on the ion collector (IC), on which they are neutralized. In the following we discuss some questions related to the measurement of isotope ratios.

Special arrangements for the *inlet system* are necessary because the instability of the ions produced and the mass separation require a high vacuum. If the mean free path length (flight without collision with other molecules) of molecules is large compared with the dimensions of the tubing through which the gas is flowing, then we refer to this as molecular flow. During molecular flow the gas particles do not influence each other. Therefore, the gas flow velocity of the lighter component is greater than the heavier component, and this means that the heavier isotope is preferentially enriched in the gas reservoir. To avoid such a mass discrimination, normally the isotope abundance measurements of gaseous substances are carried out utilizing viscous gas flow. During the viscous gas flow the free path length of molecules is small, indicating that the gas particles influence each other. The normal gas pressure is around 100 torr. At the end of the inlet system through which we have viscous gas flow, there is a "leak", an orifice or a constriction in the flow line.

In the *ion source*, where the gas pressure is between 10^{-4} to 10^{-6} torr, the gas flow is always molecular. Because ions are massive and move with a lower velocity in relation to electrons, the region through which the particle passes must be under very low gas pressure.

In general, ions are produced thermally or by electron impact. Positive ions of gaseous samples are provided most reliably by electron bombardment. The ribbon-shaped electron beam that ionizes the gas sample when it is leaked into the source is usually twisted by a coaxial magnetic field to increase the efficiency of ionization.

There is a minimum threshold energy below which ionization does not occur. The production of an excessively intense electron beam results in a large spread of velocity, and leads to poor resolution. In practice it is not sufficient for the mass analyzer to have only a dispersive property; the mass analyzer must also possess a directional focusing property. Ions emerging from the source within a small aperture angle must be refocused.

In 1940 NIER introduced the sector magnetic analyzer. In this type of analyzer, deflection takes place in a wedge-shaped magnetic field. The ion beam enters and leaves the field at right angles to the boundary, so the deflection angle is equal to the wedge angle, namely $60°$.

The sector instrument has the advantage of its source and detector being comparatively free from the mass-discriminating influence of the analyzer field.

After passing through the magnetic field, the separated ions are collected and converted into an electrical impulse, which is then fed into an amplifier. For relatively large ion currents a simple metal tube (Faraday cage) is used. It generally consists of a hollow metal tube which is grounded through a high ohmic resistor. As the ion current passes to the ground, the potential drop caused in the resistor acts as a measure of the ion current. If the ion currents are small, then some means must be provided for magnifying the current strength.

By collecting two ion beams for the isotopes in question, simultaneously and by measuring the isotope ratio directly, a much higher precision can be obtained than from a single ion beam collection. With simultaneous collection, the isotope ratios of two samples can be compared quickly under nearly identical conditions. NIER et al. (1947) developed this technique for routine measurements, which is shown, combined with one possible read-out system (after NIELSEN, 1968b) in Fig. 6. The DC feedback amplifiers A and B are coupled to the collectors for the ion beams of the main isotope (i_B) and the rare isotope (i_A). The output signal, $-U_B$, of amplifier B is fed to a precision voltage divider (Kelvin bridge type) and the fraction, p, of $-U_B$ balances the input signal obtained from ion beam i_A. Thus, with proper adjustment of p, zero signal is obtained from amplifier A. In this case,

$$U_A - pU_B = 0 \, .$$

The zero adjustment is realized for the standard only, and an output signal zero is recorded for the unknown sample. Thus, alternating the

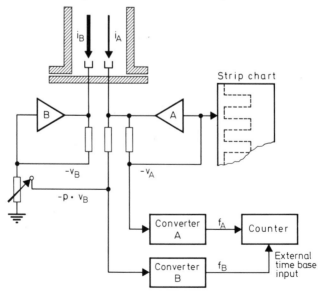

Fig. 6. Principle of digital ion beam ratio measurement. (After NIELSEN, 1968b)

inlet of the standard and unknown samples produces a step-voltage track shown in the top right of Fig. 6. After calibration, these steps give the difference, Δ, between the ion beam ratios of the standard and the unknown samples directly. From the numerical value of Δ the desired δ-value (for the definition, see below) of the sample is calculated by the equation

$$\delta X_{(sample)} = \delta X_{(standard)} + \Delta (1 + \delta X_{(standard)} + C).$$

X refers to the rare isotope in question, "standard" refers to the working standard with $\delta \neq 0$, and C is a correction for superposition of the rare isotope mass number with other molecules, e.g., $^{13}C^{16}O_2$ with $^{12}C^{16}O^{17}O$. In the analysis of carbon we collect masses 44 and 45 simultaneously, where the ratio is:

$$R = \frac{^{13}C^{16}O^{16}O + {}^{12}C^{17}O^{16}O}{^{12}C^{16}O^{16}O}.$$

The component $^{12}C^{17}O^{16}O$ constitutes about 7% of the total mass 45. The correction factor for carbon analysis contains a constant term reflecting the ^{17}O-content of the standard and a variable term reflecting the difference in the ^{17}O-content of the sample and the standard. The latter correction is obtainable directly from an ^{18}O-analysis of the gas being analyzed for ^{13}C (fractionations for various isotopes increase in proportion to the mass difference, i.e., $\delta {}^{18}O = 2\delta {}^{17}O$). For a more detailed discussion see CRAIG (1957).

The digitization equipment, giving a direct readout of Δ, is shown in the lower part of Fig. 6. The main components are two voltage-to-frequency converters and an electronic counter equipped for frequency ratio measurements.

The converters provide pulse frequencies directly proportional to the input voltages, i.e., f_B represents $-pU_B$, and f_A represents the small difference signal obtained from amplifier A in the measurement of the unknown sample. f_A is fed to the normal counter input, whereas f_B is fed to the external input of the time base decades. In this way a direct frequency ratio measurement is provided.

The main feature of this digitization method is its true integration of signal voltage during the entire counting time.

V. Standards

In stable isotope geochemistry we are concerned with measuring small changes in isotope ratios, where the isotope ratio must be measured relative to an arbitrary standard. The accepted unit of isotopic ratio measurement is the delta-value (δ), given in per mil ($\%_0$). The δ-value is defined as

$$\delta \text{ in } \%_0 = \frac{R_{(sample)} - R_{(standard)}}{R_{(standard)}} \times 1000$$

where R represents the isotope ratio. If $\delta_A > \delta_B$, we speak of A being "heavier" than B.

Unfortunately, not all of the δ-values cited in the literature, are given relative to one of the world-wide standards, so that often several standards of one element are in use. To convert δ-values from one standard to another, CRAIG (1957) has given the following equation

$$\delta_{(x-A)} = \delta_{(x-B)} + \delta_{(B-A)} + 10^{-3} \delta_{(x-B)} \delta_{(x-A)}$$

δ_{x-A} and δ_{x-B} are δ-values of sample x with relation to standards A and B; δ_{B-A} is the δ-value of B with relation to A.

For each element each laboratory uses a convenient "working standard". However, all values measured relative to the "working standard" should be calculated and given in the literature as relative to a universal standard. Unfortunately, there often has been a general lack of agreement among stable isotope researchers as to what standard should be assigned as the universal standard.

The question of which standard is the most suitable for stable isotope determinations is very complicated. The National Bureau of Standards (NBS) has undertaken a program for the preparation and distribution of isotopic reference samples. The NBS serves as a clearinghouse for data, and distributes the accumulated data with each sample requested.

A standard should fulfil the following requirements: It should be
1) used world-wide as the zero point,
2) homogeneous in composition,
3) available in relatively large amounts,

Table 3. World-wide standards in use for the isotopic composition of hydrogen, carbon, oxygen and sulfur

Element	Standard	Standard abbreviated
H	Standard Mean Ocean Water	SMOW[a]
C	Belemnitella americana from the Cretaceous PeeDee formation, South Carolina	PDB
O	Standard Mean Ocean Water	SMOW[a]
S	Troilite (FeS) from the Canyon Diablo iron meteorite	CD

[a] SMOW is defined in terms of NBS-1, a distilled water sample, such that D/H (SMOW) = 1.050 D/H (NBS-1). $^{18}O/^{16}O$ (SMOW) = 1.008 $^{18}O/^{16}O$ (NBS-1) (CRAIG, 1961b).

4) easy to handle for chemical preparation and isotopic measurement,

5) in about the medium value of the naturally occurring isotopic variation range.

Among the reference samples now used, relatively few fulfil all of the requirements. For example, the most widely used carbon isotope standard, PDB, a belemnite, is now exhausted. The NBS distributes another standard, Solenhofen limestone (NBS-20). Table 3 presents world-wide standards now in use.

VI. General Remarks on Sample Handling

Isotopic differences between the samples to be measured are often exceedingly small. Therefore, great care has to be taken in chemical handling of the sample material, to avoid any isotope fractionation there.

There are difficult chemical problems to be solved in the chemical preparation of minerals and rocks for isotopic analysis. It is important to carry out chemical reactions that convert rock samples to gas samples without changing the isotopic composition by contamination (e.g., contamination through the reagents), by exchange with extraneous matter, or by chemical isotope fractionation.

To convert the samples to a suitable form for analysis, several different preparation techniques are used in isotope geochemistry. Although these preparation procedures are chemically very different, the various methods have one general feature in common: Any preparation procedure providing a yield of less than 100 percent may produce a reaction product that is isotopically different from the original specimen, since the different isotopes exhibit different reaction rates.

A quantitative yield of a pure gas is necessary for the mass-spectrometric measurement in order to prevent not only isotope fractionations during sample preparation, but also interference in the mass spectrometer. Contamination with gases having the same molecular masses and having similar physical properties may be a serious problem. This is especially critical with CO_2 and N_2O on the one hand (CRAIG and KEELING, 1963) and N_2 and CO on the other. When CO_2 is used, interference by hydrocarbons and a CS^+ ion may also be a problem.

The chief contaminant of gaseous samples is air. Contamination may result from incomplete evacuation of the vacuum system and/or from degassing of the sample. How gases are transferred, distilled, or otherwise processed in vacuum lines is briefly discussed under the different elements. All errors due to chemical preparation commonly limit the precision to 0.1 to 0.3‰.

VII. Some General Trends in Stable Isotope Geochemistry

The foundations of stable isotope geochemistry were laid in 1947 by UREY's paper on the thermodynamic properties of isotopic substances and by NIER's development of the ratio mass spectrometer.

As we shall see, the most interesting results in stable isotope geochemistry have been established with those elements from which a stable gaseous component is known, e.g., hydrogen, carbon, oxygen, and sulfur. There are two reasons for this: The differences in isotope composition found in nature within these elements are very large, and gaseous compounds can be measured with better reproducibility than solid substances.

Before going into details of the naturally occurring variations of stable isotopes, it might be useful to discuss some general trends effective over the whole field.

1) Detectable isotope fractionations occur only when the relative mass differences between the isotopes of a certain element are relatively large, i.e., measurable isotope fractionations should be detectable only within the lightest elements (in general, up to a mass number around 40).

2) All those elements that form solid, liquid, and gaseous compounds which are stable over a wide temperature range, are likely to have variations of isotopic composition. Examples are hydrogen, carbon, nitrogen, oxygen, and sulfur.

3) With increasing temperatures the fractionation factors decrease, which means that the high-temperature environment, e.g., magmatic rocks, shows smaller fractionations than the low-temperature environment, i.e., sedimentary rocks.

4) During biological reactions, e.g., during photosynthesis, bacterial reactions, and other microbiological processes, the lighter isotope is preferentially enriched in the reaction product relative to the starting substances. This is especially pronounced in the cases of carbon, hydrogen, and nitrogen. When we consider biological processes such as photosynthesis and bacterial sulfate reduction, we must remember that isotopic variations during the history of the earth have probably increased. Therefore, during late Precambrian time the occurring variations should be smaller than at the present. Perhaps this could be proven in the case of sulfur and carbon.

The general result of all naturally occurring fractionation processes is that the degree of isotope fractionation increases with progression from the deep-seated material to materials at the surface of the earth.

B. Fractionation Mechanisms of Selected Elements

I. Hydrogen

Until 1931 it was assumed that hydrogen consisted of only one isotope. UREY *et al.* (1932a, b) established the presence of a second, heavy, stable isotope, which was called deuterium. WAY *et al.* (1950) gave the following average abundances of the stable hydrogen isotopes:

$$^1H : 99.9844\% ,$$
$$^2D : \ 0.0156\% .$$

(In addition to these two stable isotopes there is a third naturally occurring but radioactive isotope, ^3T, tritium, with a half-life of approximately 12.5 years.)

The isotope geochemistry of the stable hydrogen isotopes is one of the most interesting, for several reasons.

1) Hydrogen shows by far the largest relative mass difference between its two stable isotopes. This results in hydrogen showing the largest fractionations for stable isotopes on Earth up to 70% and even more in lunar material (see Fig. 7).

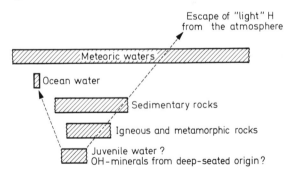

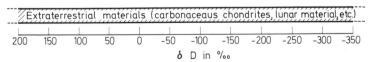

Fig. 7. D/H ratios of some important terrestrial compounds (δ-variations in ‰ relative to SMOW

2) Hydrogen is nearly omnipresent in the forms H_2O, OH, H_2-gas, and CH_4, even at great depths in the earth's mantle. Therefore, it is conceivable that hydrogen plays a major role, directly or indirectly, in almost all naturally occurring geological processes.

3) All major isotope fractionation processes are effective in the hydrogen cycle.

In Fig. 7 natural variation ranges of hydrogen isotope composition of some geologically important reservoirs are given.

1. Preparation Technique and Mass-Spectrometric Measurement

The determination of the D/H ratio is usually carried out on H_2 gas. Water is converted to hydrogen by being passed over hot metallic uranium at about 750° C e.g., as described by CRAIG (1961a) and GODFREY (1962). Most of the hydrogen generated from hydroxyl-bearing minerals is liberated in the form of water, but some is liberated as molecular hydrogen (SAVIN and EPSTEIN, 1970a). The resulting H_2 gas is converted in many laboratories to water by reaction with copper oxide. The water is then treated as described above.

Due to the great relative mass difference between H and D, the separation angle between the two isotopes is so large that a special type of mass-spectrometer with an additional tube joined to the usual tube has been developed (e.g., FRIEDMAN, 1953).

A difficulty in measuring D/H isotope ratios is that, along with the H_2^+ and HD^+ formation in the ion source, H_3^+ is produced as a by-product through the reaction:

$$H_2^+ + H \rightarrow H_3^+ \ .$$

The resolution between HD^+ and H_3^+ in most mass-spectrometers for isotope abundance measurements is not large enough to remove H_3^+; therefore, a H_3^+ correction has to be made. The H_3^+ concentration is minimized by working at low gas pressures, by addition of a repeller electrode to the source, and by use of as high an accelerating voltage as possible (FRIEDMAN, 1953). The analytical error for hydrogen isotope data is usually given as ± 1 to $\pm 2‰$.

2. Standard

The standard now used world-wide is SMOW (Standard Mean Ocean Water) (CRAIG, 1961b). D/H ratios, however, occur in the literature for at least six different "working tap water" standards.

Because there are large variations in δ-values, the deviations from the standard can be given in the literature either in % or in ‰. Like all other stable isotope data, all δD-values are given in the following in ‰.

3. Fractionation Mechanisms

a) Vapor Pressure and Freezing-Point Differences

The most effective processes that produce hydrogen isotope variations are those due to vapor pressure differences, and to a much smaller degree, those due to differences in freezing points. Because the vapor pressure and freezing point of HDO are slightly lower than those of

H_2O, the concentration of D is higher in the liquid phase than in the vapor, and higher in the solid phase than in the liquid, when water undergoes a phase change under equilibrium conditions.

The physical processes responsible for the fractionation of hydrogen isotopes in water and the distribution of the resulting fractions in nature are the same as those applying to the fractionation of oxygen isotopes in water. Therefore, the fractionation of D parallels that of ^{18}O in most cases.

b) Equilibrium Exchange Reactions

BOTTINGA (1969a) has calculated the fractionation factors for water, hydrogen, and hydrogen-methane. Large fractionations (more than 70‰ at about 350° C) occur in the system water-vapor and methane. Deuterium exchange between H_2S and H_2O is utilized commercially in the production of heavy water.

SUZUOKI and EPSTEIN (1970) determined the hydrogen isotope fractionation factors between muscovite, biotite, hornblende, and water over a temperature range of 400 to 700° C.

Table 4. Assumed values of equilibrium for mineral-H_2O isotopic fractionations. (After SUZUOKI and EPSTEIN, 1970)

Coexisting pair	Temperature, °C	D/H ratio
Biotite-H_2O	700	-20 (± 10)
	400	-50 (± 10)
Muscovite-H_2O	700	$- 5$ (± 10)
	400	-20 (± 10)

For those cations to which hydroxyl is directly attached, the substitution of Fe for Al or Mg is known to have an important effect on hydrogen isotope fractionations. For example, muscovite always concentrates D relative to Fe-bearing biotite (TAYLOR and EPSTEIN, 1966b), whilst Mg-rich minerals such as phlogopite do the same.

Data on approximate fractionation factors for clay mineral-water and hydroxides-water systems at temperatures of sedimentary formation were deduced from the isotopic compositions of natural samples by SAVIN and EPSTEIN (1970a) and LAWRENCE and TAYLOR (1971).

Mineral	Fractionation factor α for mineral-water
Kaolinite and montmorillonite	0.97
Glauconite	0.93
Gibbsite	0.985

FONTES and GONFIANTINI (1967) determined a decrease of 15‰ in deuterium in the chemically bound water of gypsum, which is in equilibrium with ocean water.

Equilibrium has not been proved in all of the above cases, but these examples of chemical exchange reactions demonstrate how many possibilities exist for hydrogen isotope fractionations.

c) Kinetic Isotope Effects

Appreciable hydrogen isotope fractionation seems possible in biochemical processes, e.g., during photosynthesis (KRICHEVSKY et al., 1961) and during bacterial production of molecular hydrogen and methane.

II. Carbon

1. Introductory Remarks

Carbon is one of the most abundant elements in the universe, but it occurs in the earth as a trace element. The average carbon content of the crust and the mantle probably lies in the range of several hundred parts per million. Besides playing a key role in the biosphere, inorganic carbon is important because it shows a diversity of components, e.g., diamond, carbonates, carbon dioxide. This distribution of more oxidized carbon compounds of inorganic origin and of more reduced carbon in the biosphere is an ideal situation for the search for naturally occurring isotope fractionations.

Carbon has two stable isotopes:

$^{12}C = 98.89\%$ (reference mass in modern compilations)

$^{13}C = 1.11\%$ (NIER, 1950).

The naturally occurring variations of the carbon isotope composition is greater than 10% (neglecting meteoritic carbonate). Heavy carbonates with δ-values of more than $+20$ and light methane of around $-90‰$ have been reported in the literature (see also Fig. 9).

Since the very first measurements of NIER and GULBRANSEN (1939) and MURPHEY and NIER (1941), we know that the carbonates concentrate ^{13}C relative to organic compounds.

A great many $^{13}C/^{12}C$ mass-spectrometric determinations have been made in the last 20 years; many of these are unpublished because they are of commercial interest for petroleum industries. The first — and one of the best — summaries of the carbon isotope variations of the different carbon compounds has been given by CRAIG (1953). Two recent compilations have been written by SCHWARCZ (1969) and DEGENS (1969).

2. Preparation Technique

The gas used in all $^{13}C/^{12}C$ measurements is CO_2, for which the following preparation methods exist:

1) *Carbonates* react with 100% pure phosphoric acid at room temperature (25° C, McCREA, 1950) to liberate CO_2. Some carbonates, like siderite, react very slowly with H_3PO_4, and the yields after a 24-hour period are much below the theoretical maximum. In the case of siderite, WEBER *et al.* (1964) argued that after 20 min of reaction, the $^{13}C/^{12}C$ ratio lies within the experimental error range with the theoretical value.

2) *Organic compounds* are generally oxidized in a stream of oxygen at temperatures between 900 and 1000° C. In addition to CO_2, CO is liberated and must be completely oxidized to CO_2 over a catalyst or through a "sparking technique" (CRAIG, 1953). Another technique has been used by HOEFS and SCHIDLOWSKI (1967), who mixed organic compounds, with V_2O_5 as the oxidizing substance, and reacted the mixture at 1 000° C.

3. Standards

Several standards have been used by various authors. The PDB standard in Table 5 is commonly used as the reference standard, although it has been exhausted for several years.

Table 5. Carbon isotope standards commonly used in the literature. Most δ-values are given after CRAIG (1957)

	$\delta^{13}C$
1) Chicago PDB standard (belemnitella americana from the Cretaceous Pedee formation, South Carolina)	0
2) Solenhofen limestone, NBS No. 20 (not very suitable as a standard, because it is isotopically inhomogeneous (BAUSCH and HOEFS, unpublished data)	$- 1.1$
3) $BaCO_3$, Stockholm standard	$- 10.3$
4) Graphite, NBS No. 21	$- 27.8$
5) Petroleum standard, NBS No. 22	$- 29.4$

4. Fractionation Mechanisms of Carbon Isotopes

The two main carbon reservoirs, the biosphere and the carbonates, are isotopically separated from each other through two different reaction mechanisms:

1) a kinetic effect during photosynthesis, leading to a depletion of ^{13}C in CO_2, and concentrating the light ^{12}C in the synthesized organic material,

2) a chemical exchange effect in the system: atmospheric CO_2- dissolved HCO_3^- -$CaCO_3$, in which an enrichment of ^{13}C takes place in the bicarbonate.

1 a) Metabolic Effects. The exclusive carbon source during photosynthesis seems to be molecular CO_2. Experiments carried out by PARK and EPSTEIN (1960) (see also ABELSON and HOERING, 1961) on the extraction of CO_2 and on the fixation of CO_2 by enzymes suggest that the major fractionation in the photosynthetic fixation of CO_2 occurs in two steps.

I. The first step involves the preferential intake of $^{12}CO_2$ from the atmosphere. The fractionation accompanying the first step can vary over a wide range.

II. In the second step $^{12}CO_2$ is preferentially fixed into the primary photosynthetic product (known as "phosphoglyceric acid"), which is ultimately converted to carbohydrate. If all the dissolved CO_2 were fixed, no difference in the $^{13}C/^{12}C$ ratio would exist between the CO_2 and the carbohydrate produced. Therefore, it is postulated that a portion of the unreacted and ^{13}C-enriched, dissolved CO_2 has to be eliminated from the plant. The more efficient the elimination is, the greater the fractionation in the second step.

1 b) Environmental Effects. DEGENS et al. (1968a, b) concluded from experiments with marine plankton that the fractionation is principally controlled by pH, water temperature, concentration of carbonic acid species, and growth rates of organisms. As a general trend, maximum fractionation is achieved when pH and water temperature are low, the dissolved CO_2 concentration is high, and the growth rate is moderate.

In this connection it is interesting to note that UPHAUS and KATZ (1967) have demonstrated that plants grown in D_2O show a decreased tendency to fractionate ^{13}C during photosynthesis. The change in fractionation shown by deuterated organisms is most likely a consequence of widespread changes in the cell. KATZ and CRESPI (1966) have demonstrated that deuterated plants show significant differences in morphology and biochemistry as compared to the same species of normal isotopic composition.

2) Isotope Equilibrium between Gaseous CO_2 and Dissolved Bicarbonate. The equilibrium fractionation in the exchange reaction

$$^{13}CO_2 + H^{12}CO_3^- \rightleftharpoons {}^{12}CO_2 + H^{13}CO_3^-$$

has been determined several times (Table 6).

The ratio of the carbonic acid substances in solution is pH- and temperature-controlled. In ocean water (pH = 8.2) more than 99% of its dissolved inorganic carbon is HCO_3^-; in more acidic waters, such as some fresh waters, molecular CO_2 becomes an important species. This in turn must control the carbon isotope composition[1].

---

1 Observed differences in $^{13}C/^{12}C$ ratios between fresh waters (light) and marine waters (heavy) cannot be exclusively interpreted as a result of a biogenic contribution to rivers and lakes.

Table 6. Experimentally determined carbon isotope fractionations between dissolved bicarbonate and gaseous CO_2 (the values are calculated for 20° C, using a temperature dependence of $-0.109‰/°$ C)

Author	Fractionation in ‰ (20° C)
VOGEL (1961)	7.9
ABELSON and HOERING (1961)	7.8
DEUSER and DEGENS (1967)	7.6
WENDT (1968)	8.2
EMRICH et al. (1970)	8.4

3) Other important processes producing isotope fractionations in the carbon cycle on a wide scale are:

a) A kinetic effect arises during the thermal cracking of hydrocarbons. FRANK and SACKETT (1969) have found a ^{13}C depletion in methane generated during the thermal cracking of neopentane. This suggests that

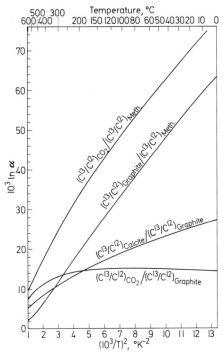

Fig. 8. Calculated carbon isotope fractionation factors for exchange among graphite, calcite, carbon dioxide, and methane. (After BOTTINGA, 1969a)

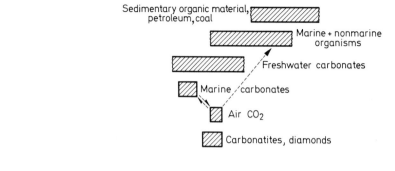

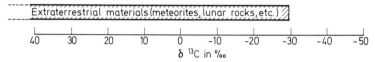

Fig. 9. $^{13}C/^{12}C$ ratios of some important carbon compounds (δ-variations in ‰ relative to PDB)

cracking reactions are extremely important for the isotopic composition of light hydrocarbons, especially methane.

b) BOTTINGA (1969a) has calculated the fractionation factors for the isotope-exchange systems involving pairings of calcite, carbon dioxide, graphite, and methane. Second to the calcite-carbon dioxide system, the carbon dioxide-methane system is of great geological importance (see Fig. 8).

c) Other phenomena producing isotope fractionations are diffusion processes, for instance, during the migration of methane and other light hydrocarbons. In this connection adsorption-desorption phenomena may also influence the carbon isotope ratios of methane and related substances.

In Fig. 9, $\delta^{13}C$-variations of some important carbon compounds are schematically demonstrated.

III. Oxygen

Oxygen is the most abundant element on earth. It occurs in gaseous, liquid, and solid compounds, most of which are thermally stable over large temperature ranges. These facts make oxygen one of the most interesting elements in isotope geochemistry. A large amount of literature deals with the naturally occurring variations of oxygen isotopes. There have been several review papers on this subject: SILVERMAN

(1951), EPSTEIN (1959), CLAYTON (1963a), EPSTEIN and TAYLOR (1967), TAYLOR (1968a), and GARLICK (1969).

Oxygen has three stable isotopes with the following abundances (GARLICK, 1969):

$$^{16}O = 99.763 \ \% \ ,$$
$$^{17}O = \ \ 0.0375\% \ ,$$
$$^{18}O = \ \ 0.1995\% \ .$$

Because of the higher abundance and the greater mass difference, the $^{18}O/^{16}O$ ratio is normally determined.

1. Preparation Techniques

In almost all cases, CO_2 is the gas used in the mass-spectrometric measurement. Different methods are necessary to liberate the oxygen from the various oxygen-containing compounds.

The oxygen in *silicates* and *oxides* is usually converted to CO_2 through fluorination with F_2 or BrF_5 in nickel-tubes at 500 to 600° C. Decomposition by carbon reduction at 1000 to 2000° C is suitable only for quartz and iron oxides but not for silicates (CLAYTON and EPSTEIN, 1958).

The liberation of oxygen through F_2 or BrF_5 is described by TAYLOR and EPSTEIN (1962a) and CLAYTON and MAYEDA (1963). Unwanted product gases are removed from the oxygen by cold traps, and excess F_2 is removed by reaction with hot KBr to KF and condensable Br_2. The oxygen is combusted to CO_2 over a resistance-heated graphite rod.

Care must be taken to ensure quantitative oxygen yields. Low yields caused by inadequate reaction temperatures or reaction times result in anomalously low $^{18}O/^{16}O$ ratios.

Phosphates may be treated as above with a modified technique described by TUDGE (1960).

Sulfates are precipitated as $BaSO_4$, and then reduced at 1000° C with carbon to CO_2 and CO. The CO is converted to CO_2 by sparking between platinum electrodes (LONGINELLI and CRAIG, 1967).

Carbonates are reacted with 100% phosphoric acid at 25° C, first described by McCREA (1950). The following reaction scheme

$$3CaCO_3 + 2H_3(PO_4) \rightleftharpoons 3CO_2 + 3H_2O + Ca_3(PO_4)_2$$

shows that only two-thirds of the oxygen originally present in the carbonates is liberated. [This must be taken into account, when comparing δ-values given relative to the PDB with the ocean water standard (SMOW)].

There are characteristic differences in the isotopic fractionation factors associated with the phosphoric acid liberation of CO_2 from various carbonates, which has to be considered when the isotopic composition of different carbonates are compared.

Table 7. Isotopic fractionation factor of various carbonates. (After SHARMA and
CLAYTON, 1965, in GARLICK, 1969)

Mineral	α	$1000 \ln \alpha(\delta_{(CO_2)} - \delta_{(carbonate)})$
Rhodochrosite	1.01012	10.07
Calcite	1.01025	10.20
Aragonite	1.01034	10.29
Dolomite (ordered)	1.01109	11.03

Water is usually analyzed by equilibration of a small amount of CO_2 with a surplus of water. The equilibrium fractionation factor between CO_2 and H_2O has been determined to be 1.0407 at 25° C (COMPSTON and EPSTEIN, 1958; O'NEIL and EPSTEIN, 1966a). It is not necessary to know the oxygen isotopic composition of the CO_2 before equilibration with the surplus water, since the amount of oxygen in the CO_2 may be neglected.

2. Standards

The standard now used world-wide is Standard Mean Ocean Water (SMOW). The original standard, at least for the paleotemperature determinations, was the CO_2 liberated by H_3PO_4 from PDB, a Cretaceous belemnite from the Pedee formation, which has been exhausted. The total carbonate of PDB has a $\delta^{18}O$-value of $+30.4‰$ on the SMOW scale. (Note that the liberated CO_2 has an isotopic composition of $+40.9‰$. Note also that due to the different fractionation factors associated with the acid liberation of CO_2:

$$(\delta^{18}O_{(dolomite)} - \delta^{18}O_{(calcite)} \text{ SMOW}) \neq (\delta^{18}O_{(dolomite)} - \delta^{18}O_{(calcite)}) \text{ PDB} .$$

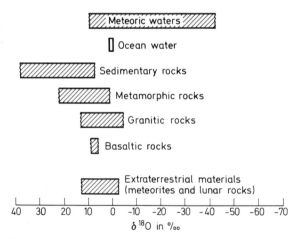

Fig. 10. $^{18}O/^{16}O$ ratios of some important oxygen-containing compounds (δ-variations in ‰ relative to SMOW)

The National Bureau of Standards now distributes another carbonate: NBS No.20. This is Solenhofen limestone, which has a δ-value of $-4.14‰$ relative to PDB (CRAIG, 1957). As shown by BAUSCH and HOEFS (unpublished data), the Solenhofen limestone is rather inhomogeneous within a single sedimentary layer. The differences in δ-values are as much as 2.5‰ in ^{18}O. Therefore, a better standard is urgently required. All data given in the following are reported relative to SMOW.

3. Fractionation Mechanisms

The $^{18}O/^{16}O$ ratio in nature exhibits variations ranging up to about 10%. These variations result from both equilibrium and kinetic fractionation effects. Fig.10 presents a schematic diagram on the naturally occurring variations of oxygen isotopes.

a) Equilibrium Exchange Reactions

An excellent consistency in the relative ^{18}O contents of different minerals can be found in nature. GARLICK (1969) made an attempt to arrange coexisting minerals according to their relative tendencies to concentrate ^{18}O.

quartz, dolomite (anhydrite)
alkali feldspar, calcite, aragonite
leucite
muscovite, nepheline
anorthite (kyanite)
glaucophane (staurolite)
lawsonite
garnet, common pyroxenes, and amphiboles
biotite
olivine (sphene)
chlorite
ilmenite (rutile)
magnetite (hematite)
pyrochlore

This order of decreasing ^{18}O-content is due to a crystal-chemical relationship associated with the relative affinity for ^{18}O. TAYLOR and EPSTEIN (1962b) have pointed out that those minerals in which the Si-O-Si bond predominates are richest in ^{18}O. At igneous temperatures the Si-O-Al types contain approximately 4‰ less ^{18}O, and the Si-O-Mg and Si-O-Fe types contain 2‰ less ^{18}O. The OH-bearing minerals have abnormally low $^{18}O/^{16}O$ ratios, suggesting that the OH-group may be lower in ^{18}O than the rest of the oxygen in the silicate structure.

It is reasonable to assume that due to the described consistency and to other observations discussed below, oxygen isotope equilibrium is not always attained in a given assemblage. However, in many cases there is a close approach to isotopic equilibrium. On the basis of these systematic tendencies in the $^{18}O/^{16}O$ ratios of many minerals, it has become apparent that significant temperature information could be obtained up to temperatures of $1000°$ C and even higher in natural assemblages, if calibration curves could be worked out for the various mineral pairs. In an assemblage of n-phases, we can obtain $n-1$ independent "temperatures of formation", one temperature for each mineral pair. Barring coincidence, if each mineral pair gives the same temperature, we can be certain that oxygen isotope equilibrium was attained in the mineral assemblage, and that equilibrium was "frozen in" at the same temperature in every mineral of the assemblage. If one mineral introduces inconsistent temperatures of formation among a set of concordant equilibrium pairs, it is certain either that it was not equilibrated with the rest of the assemblage, or that it has undergone oxygen exchange subsequent to the formation of the original assemblage. In order for a mineral pair to serve as a good isotope geothermometer, SCHWARCZ et al. (1970) gave the following criteria:

  1) they should be abundant and commonly found together,

  2) they should be stable together over a wide range of pressures and temperatures,

  3) they should exhibit limited compositional variation, since isotopic fractionation between A and B depends on the chemical composition of A and B.

Compared with other geothermometers, the advantage of isotope thermometers is to be not pressure-sensitive. The basic reason for this independence of pressure changes is that the volumes of isotopically substituted molecules and crystals do not differ appreciably.

It is important to note that the oxygen isotopic composition of a mineral is dependent upon the chemical composition of the mineral. This was for instance experimentally demonstrated for the plagioclase series by O'NEIL and TAYLOR (1967), who showed that the plagioclase-H_2O fractionation depends strongly upon the anorthite content of the plagioclase (see Fig. 11). This method has been called "internal thermometry" (ONUMA et al. 1972b). On the other hand, the paleotemperature method of UREY et al. (1951) is based on the fractionation of oxygen isotopes between calcium carbonate and seawater, so that a temperature based on isotopic measurement of fossil shells requires an additional assumption about the isotope ratio in the ocean water at the time of precipitation. This method, in which one of the phases is not available for analysis, has been called "external thermometry" (ONUMA *et al.*, 1972b).

In recent years some laboratory calibration curves have been worked out. The starting materials are chosen so that in one set of experiments the mineral approaches equilibrium by going through successively more ^{18}O-rich compositions, while in another set of experiments the reverse approach is taken. These two approaches are not always practical because laboratory exchange rates between most minerals and aqueous solutions are slow, even at high temperatures.

The calcite-water fractionation curve was the first isotopic geothermometer to be developed by CLAYTON (1961). A quartz-water calibration curve was determined by O'NEIL and CLAYTON (1964), but this has since been modified in the low-temperature range by CLAYTON *et al.* (1972). O'NEIL and CLAYTON (1964) had to rely on direct synthesis of quartz from silica gel in aqueous solutions for data points below 600° C because exchange rates between quartz and water are so slow. CLAYTON *et al.* (1972) were able to demonstrate equilibrium by using a modified two-directional exchange procedure at these lower temperatures. This involves carrying out partial exchange experiments and then extrapolating the isotopic results to equilibrium values. O'NEIL and TAYLOR (1967) found that Na$^+$ and K$^+$ exchange between alkali feldspars and aqueous chloride solutions produces essentially complete oxygen isotopic exchange in the feldspars. With this procedure they were able to use the

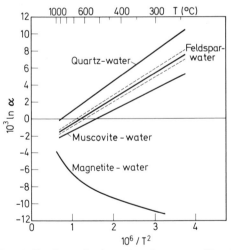

Fig. 11. Experimentally determined oxygen isotope calibration curves; quartz-water (O'NEIL and CLAYTON, 1964; CLAYTON *et al.*, 1972), feldspar-water (O'NEIL and TAYLOR, 1967), muscovite-water (O'NEIL and TAYLOR, 1966), and magnetite-water (BERTENRATH *et al.*, 1972). (After FRIEDRICHSEN, 1971)

two-directional exchange method at temperatures above 350° C to obtain an alkali-feldspar-water calibration curve. O'NEIL and TAYLOR (1966) also used the alkali ion exchange technique to obtain the muscovite-water calibration curve. These experiments were made by transforming kaolinite or paragonite to muscovite, and in both instances complete oxygen isotope exchange accompanies the cation exchange.

The most suitable and sensitive mineral pair is the magnetite-quartz pair. BERTENRATH et al. (1972) have redetermined the magnetite-water curve, revising the curve given by O'NEIL and CLAYTON (1964).

Some of the experimentally determined calibration curves are shown in Fig. 11.

b) Fractionations Due to Kinetic Processes

Oxygen isotope fractionations occurring during photosynthesis (DOLE and JENKS, 1944) and respiration (LANE and DOLE, 1956) are kinetic processes. MIZUTANI and RAFTER (1969 b) have shown that in the bacterial reduction of sulfate, at any stage of reduction the remaining sulfate is enriched not only in ^{34}S, but also in ^{18}O. The ratio of the ^{34}S enrichment to the ^{18}O enrichment has been found to be approximately 4:1.

c) Fractionations Due to Vapor Pressure Differences

Variations found in the isotopic composition of natural waters are due to vapor pressure differences. The light isotopic component has a higher vapor pressure than the heavy one, as has already been discussed.

IV. Sulfur

Sulfur has four stable isotopes with the following abundances (MACNAMARA and THODE, 1950):

$$^{32}S: \quad 95.02\% ,$$
$$^{33}S: \quad 0.75\% ,$$
$$^{34}S: \quad 4.21\% ,$$
$$^{36}S: \quad 0.02\% .$$

Sulfur is present in nearly all natural environments: As a minor component in igneous and metamorphic rocks, mostly as sulfides; in the biosphere and related organic substances, like crude oil and coal; in ocean water as sulfate and in marine sediments as both sulfide and sulfate. It may be a major component: In ore deposits, where it is the dominant nonmetal, and as sulfate in evaporites. This occurrence covers the whole temperature range of geological interest. Sulfur is bound in various valence states, from negatively charged sulfides, elemental sulfur, to positively charged sulfates. From these facts it is quite clear that sulfur is of special interest in stable isotope geochemistry.

THODE et al. (1949) and TROFIMOV (1949) were the first to observe wide variations in the abundances of sulfur isotopes.

Today variations on the order of 15% have been found, e.g., the "heaviest" sulfates show δ^{34}S-values greater than $+90\permil$ (NIELSEN, 1972a) and the "lightest" sulfides have δ-values of around $-50\permil$. For a schematic diagram, see Fig. 12.

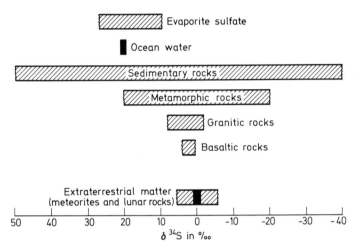

Fig. 12. ^{34}S/^{32}S distribution in some naturally occurring sulfur compounds (δ-variations in ‰ relative to Canyon Diablo troilite)

The following papers have summarized the whole field or broader aspects of the naturally occurring variations: AULT (1959), AULT and KULP (1959), STANTON (1960), THODE et al. (1961), THODE (1963, 1972), JENSEN (1967).

1. Standard

The reference standard commonly used is sulfur from troilite of the Canyon Diablo iron meteorite, with a ^{32}S/^{34}S ratio of 22.22.

2. Preparation Technique

The gas used in the mass-spectrometric measurement is mostly SO_2. PUCHELT et al. (1971b) described a method using SF_6. Some aspects concerning the chemical preparation of the various sulfur compounds have been discussed by RAFTER (1957) and RICKE (1964). Pure sulfides are converted to SO_2 by reaction with an oxidizing agent, like CuO,

V_2O_5 and O_2. In the case of V_2O_5, pure sulfides are mixed with V_2O_5 and sealed in a quartz tube, which is heated up to $1\,000°$ C. At this temperature, V_2O_5 releases sufficient oxygen to oxidize the sulfide to SO_2. To avoid the formation of SO_3, which would result in an isotope fractionation between SO_2 and SO_3, the quartz tubes are rapidly cooled to room temperatures.

For the extraction of sulfates and total sulfur a suitable solvent and reduction reagent is needed. THODE et al. (1961) used a reducing agent which was a mixture of HCl, H_3PO_2 and HI. The H_2S that forms may be precipitated as CdS.

Sulfides disseminated in rock samples are disintegrated with hydrochloric acid in the presence of aluminum according to the method described by RICKE (1964). The resulting H_2S may be precipitated as CdS. For pyrite, however, another preparation technique is necessary, because pyrite is attacked too slowly with HCl. Pyrite may be oxidized with nitric acid and bromine, and the resulting sulfate is then treated as above.

3. Fractionation Mechanisms

There are two types of reactions producing sulfur isotope variations:

1) a kinetic effect, during the bacterial reduction of sulfate to "light" H_2S, which gives by far the largest fractionations in the sulfur cycle

2) various chemical exchange reactions, e.g., between sulfate and sulfides on the one hand, and between the sulfides themselves on the other, where there is a definite order of concentrating ^{34}S.

1) There are certain species of bacteria, of which the best known is Desulphovibrio desulphuricans, living under anaerobic conditions, which reduce sulfate into "light" H_2S. THODE et al. (1951), JONES and STARKEY (1957), FEELY and KULP (1957), HARRISON and THODE (1957a, b), KAPLAN et al. (1960), NAKAI and JENSEN (1960 and 1964), KAPLAN and RITTENBERG (1964), and KEMP and THODE (1968) showed that fractionations of up to nearly 5% could be achieved in the laboratory under a variety of conditions.

In the experiments cited above such parameters as temperature, hydrogen donor type and concentration, sulfate concentration, and bacterial population density were varied.

KEMP and THODE (1968) argued that the following steps are involved in the reduction process:

    a) intake of sulfate,
    b) organic complexing of sulfate,
    c) reduction of sulfate to sulfite,
    d) reduction of organically bound sulfite,
    e) production of hydrogen sulfide.

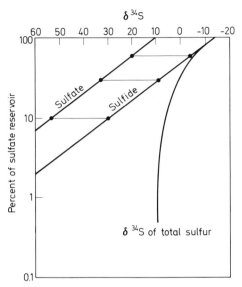

Fig. 13. Variations of ^{34}S values of sulfide produced and of residual sulfate in a closed system. Assumed fractionation factor: 1.025; assumed starting composition of sulfate: $+10$; assumed starting composition of sulfide: -15

Under different conditions the net isotope fractionation produced will depend on the isotope effects in each of these steps, the relative speeds of the steps, and the extent to which the sulfate reservoir is depleted.

Sulfate is reduced to sulfide in nature in two distinct situations, open and closed systems. In the open system the sulfate-reducing bacteria are in good contact with the infinite reservoir, and the sulfate that they reduce is constantly replenished. In such circumstances the sulfide produced will have a constant δ^{34}S-value relative to the sulfate reservoir. In a closed system the bacteria are in contact with only a limited amount of sulfate. In such cases the δ^{34}S-values of the sulfide produced and of the residual sulfate depend on the extent of reaction of the available sulfate reservoir (schematically shown in Fig. 13).

2) The isotope exchange between sulfate and sulfide may be written:

$$^{32}SO_4^{--} + H_2^{34}S \rightleftharpoons {}^{34}SO_4^{--} + H_2^{34}S .$$

The theoretical value of the exchange constant is 1.075 at 25° C (TUDGE and THODE, 1950). Therefore, if this exchange takes place, although no mechanism is known yet, it should lead to sulfides being depleted in ^{34}S by amounts up to 75‰ relative to sulfates. SAKAI (1957) has extended this

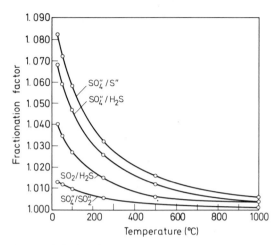

Fig. 14. Variation of the fractionation factors of sulfur compounds with temperature. (After SAKAI, 1957)

calculation up to temperatures of 1 000° C. In Fig. 14 the variation of fractionation factors with temperature is shown for some sulfur exchange reactions. THODE *et al.* (1971) have measured the equilibrium constant K between SO_2 and H_2S in the temperature range 500 to 1 000° C and compared the results with theoretical calculations.

SAKAI (1957) first suggested that isotopic fractionation between different metallic sulfides would bring about a slight variation in their isotope ratios during their deposition. Theoretical studies of fractionations between sulfides have been done by SAKAI (1968) and BACHINSKI (1969), who reported the bond strength of sulfide minerals and their relationship to isotope fractionation.

SAKAI (1968) predicted the extent of fractionation among the three common minerals pyrite, sphalerite, and galena. It should be emphasized in this connection that the main parameters influencing the isotope fractionation are the nature of the predominant sulfur species in the ore fluid, the nature of the sulfide minerals, and the temperature. OHMOTO (1972) has extended these considerations and demonstrated the influence of the fugacity of oxygen and pH of the ore-forming fluids for the sulfur isotope composition. Fractionation curves based on the estimates of SAKAI (1968) are shown in Fig. 15.

4. Hydrothermal Sulfur Isotope Distribution Experiments

During the last 20 years of sulfur isotope determinations, isotope fractionation due to biological processes has received much attention,

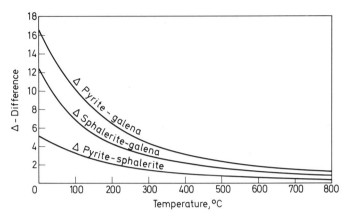

Fig. 15. Theoretical sulfur isotope fractionation curves for pyrite-galena, pyrite-sphalerite, and sphalerite-galena based on SAKAI's (1968) estimates of partition function ratios for crystalline FeS_2, PbS, and ZnS. (After BACHINSKI, 1969)

but comparatively few efforts have been made to determine sulfur isotope fractionation due to inorganic reactions, e.g., in metallic sulfides under controlled conditions. However, in the last few years more and more data of experimental determinations of the distribution of sulfur isotopes in sulfides under "hydrothermal" conditions have become available.

Since we know that in many types of mineral deposits a regular trend of isotopic fractionation is observed among coexisting sulfide minerals (the enrichment of ^{34}S is usually pyrite > sphalerite > chalcopyrite > galena) and since this relationship was confirmed from theoretical treatment by SAKAI (1968) and BACHINSKI (1969), there have been several attempts to make this confirmation from the experimental side [GROOTENBOER and SCHWARCZ (1969), KAJIWARA et al. (1969), RYE and CZAMANSKE (1969), PUCHELT and KULLERUD (1970), SALOMONS (1971), SCHILLER et al. (1970), and KAJIWARA and KROUSE (1971)].

Two essentially similar approaches have been used. One approach is to have both sulfides present in the equilibration vessel but to keep them physically separated and to effect isotope exchange between them via transport of sulfur vapor. The second approach is if hydrothermal solutions are to be used instead of a gas phase. The problem involved in this technique comes from the difficulty in reacting both mineral species of the pair under identical pH conditions.

The experimental determinations of equilibrium constants performed so far do not agree very well with each other. Before such determinations can be used as satisfactory calibrations of geothermometers, it is clearly necessary that the discrepancies be removed.

V. Selenium

Since selenium is, to some extent, chemically similar to sulfur, one might expect to find some analogous fractionations of the selenium isotopes in nature.

Six stable selenium isotopes are known with the following abundances (BAINBRIDGE and NIER, 1950):

$$^{74}Se: \qquad 0.87\% ,$$
$$^{76}Se: \qquad 9.02\% ,$$
$$^{77}Se: \qquad 7.58\% ,$$
$$^{78}Se: \qquad 23.52\% ,$$
$$^{80}Se: \qquad 49.82\% ,$$
$$^{82}Se: \qquad 9.19\% .$$

$^{82}Se/^{76}Se$ ratios have been determined by KROUSE and THODE (1962). Selenium was extracted in its elemental form from natural samples and was then fluorinated to SeF_6 and introduced into the mass-spectrometer.

From theoretical calculations KROUSE and THODE (1962) deduced that ^{76}Se and ^{82}Se differ in their chemical properties to the extent that isotope fractionations up to 6% are predicted for $^{76}Se/^{82}Se$ exchange processes, provided that mechanisms are available for such exchanges.

The $^{82}Se/^{76}Se$ ratio of 16 natural samples analyzed by KROUSE and THODE (1962), varied by 15‰ (Fig. 16). In this connection it is interesting to note that in addition to sulfate-reducing bacteria, anaerobic bacteria

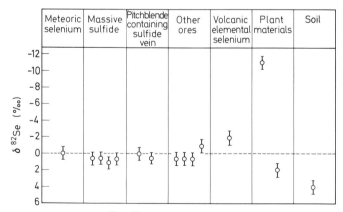

Fig. 16. Variations in the $^{82}Se/^{76}Se$ ratio found for natural samples. (After KROUSE and THODE, 1962)

are known to reduce selenates and selenites. A kinetic isotope effect of 15‰ has been found in the reduction of selenite ion to elemental selenium. Summarizing, we may state that selenium isotope fractionations should be similar to those found for sulfur.

VI. Boron

Boron has two stable isotopes (BAINBRIDGE and NIER, 1950):

$$^{10}B: \quad 18.98\% ,$$
$$^{11}B: \quad 81.02\% .$$

Boron is known to be a highly mobile element geochemically and therefore may well reveal variations in its isotopic composition. For example, boron minerals of pegmatites and hydrothermal veins (tourmaline) on one side, and hydrated borate minerals (borax) on the other side, might display isotope fractionation. The possible fractionations in biological systems should also be seriously considered. UREY (1947) has calculated the equilibrium constants of boron halogenides which differ from unity.

The corresponding data on boron isotope variations partly contradict each other: PARWEL et al. (1956), MELTON et al. (1956), LEHMANN and SHAPPIRO (1959), FINLEY et al. (1962), SHIMA (1963), MCMULLEN et al. (1961), SHERGINA et al. (1968), SCHWARCZ et al. (1969) and AGYEI and MCMULLEN (1968).

These contradicting results might be due mainly to erroneous measurements, because boron is difficult to handle in mass-spectrometric measurements. This occurs when boron is used both as a gaseous compound (BF$_3$ may cause memory effects, MELTON et al. (1956)) and as a solid compound (inherent errors in absolute ratio determinations).

Recent investigations by AGYEI and MCMULLEN (data to be published) suggest that the $^{11}B/^{10}B$ ratios in terrestrial materials fell within a narrow range of around 4.0, whereas the seawater value fell far outside this range (see Fig. 17).

Assuming that the $^{11}B/^{10}B$ ratio of about 4.0 is the mean composition of boron being fed into the oceans by erosion and volcanic eruption, SCHWARCZ et al. (1969) were surprised to find that the mean isotopic composition of ocean water is approximately 5% enriched in ^{11}B with respect to its presumed source. To explain this observation, SCHWARCZ et al. (1969) suggested that light boron is preferentially incorporated into clay minerals, especially illite.

SCHWARCZ et al. (1969) tested this hypothesis using illite and have found a fractionation of about 3 to 4% enriching the clay in ^{10}B. This is in approximate agreement with a calculated value, assuming that the ocean is at an isotopically steady state.

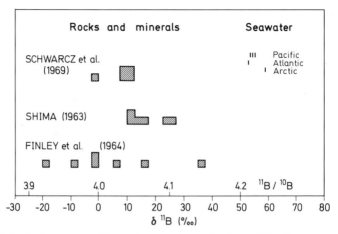

Fig. 17. Variation ranges of boron isotope determinations. (After SCHWARCZ *et al.*, 1969)

VII. Nitrogen

Nitrogen consists of two stable isotopes, ^{14}N and ^{15}N. Atmospheric nitrogen, as determined by NIER (1950), has the following composition:

$$^{14}N: \quad 99.64\% ,$$
$$^{15}N: \quad 0.36\% .$$

The various naturally occuring nitrogen compounds are prerequisite for interesting stable isotope research, the most important supposition being the existence of different gaseous components in various oxidation states.

The existence of variations in the isotopic composition of nitrogen (see Fig. 18) has been confirmed by several authors: HOERING (1955,

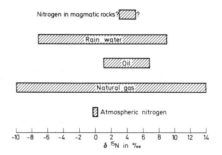

Fig. 18. $^{15}N/^{14}N$ ratios of some nitrogen-containing compounds (δ-variations in ‰ relative to atmospheric nitrogen)

1956, 1957), MAYNE (1957), PARWEL *et al.* (1957), HOERING and MOORE (1958).

Nitrogen isotope ratios are measured in the form of N_2. The standard gas used is atmospheric nitrogen.

Different preparation techniques are used to convert organic nitrogen, inorganic nitrates, and nitrogen from rocks and minerals. It is imperative that N_2 be free of carbon monoxide, which gives an interfering mass-spectrum.

Very little can be said about the fractionation mechanisms that might be responsible for the variations in the $^{15}N/^{14}N$ ratios. Only a few equilibrium processes appear to fractionate the nitrogen isotopes. It is well-known that biological processes play the key role in the nitrogen cycle, but the extent to which the abundance of ^{15}N relative to ^{14}N may be altered is not known. WELLMANN *et al.* (1968) have studied the behavior of several microorganisms participating in the nitrogen cycle. In all experiments, $^{14}NO_3^-$ and $^{14}NO_2^-$ have been preferentially reduced.

MAYNE (1957) and SCALAN (1957) have analyzed the isotopic composition of magmatic nitrogen, present in trace amounts in all igneous rocks (WLOTZKA, 1961). MAYNE (1957) found large variations in igneous rocks from -15.8 to $30.9‰$. He assumed he had found a correlation between the nitrogen content and its isotopic composition in which the more depleted the rock is in nitrogen, the more enriched it is in ^{15}N.

SCALAN (1957), however, reports a very narrow range in isotopic composition from $+3.0$ to $+5.0‰$ in igneous rocks.

Small variations exist in biogenic matter, from a given locality and even among the various parts of given species. CHENG *et al.* (1964) studied the isotope abundance in different soils and in various forms of soil nitrogen. The $^{15}N/^{14}N$ isotope variation appears to be small.

HOERING and MOORE (1958) and EICHMANN (1969) have measured a number of N_2 samples from natural gas and crude oil and found wide variations. The origin of the N_2 is not known, but several sources are possible. It may have originated from the decomposition of the nitrogenous compounds in crude oils or sedimentary material. This is suggested by the fact that the nitrogen in natural gas is lighter than the nitrogen in the accompanying crude oil. (The molecules with the lighter isotope would decompose more readily, enriching the gas phase in ^{14}N relative to the remaining oil.) Entrapped air may also be a source in some cases (EICHMANN, 1969).

In summary, nitrogen isotope determinations seem to be a fruitful field, but much more data are needed to obtain a better understanding of the naturally occurring variations.

VIII. Silicon

Silicon has three stable isotopes, with the following abundances (BAINBRIDGE and NIER, 1950):

$$^{28}Si: \quad 92.27\% \,,$$
$$^{29}Si: \quad 4.68\% \,,$$
$$^{30}Si: \quad 3.05\% \,.$$

Silicates are stable over the whole temperature range and are of interest in stable isotope research. We might expect small fractionations, small, because no geologically important gaseous silicon compound is known and because the liquid components are of minor importance. Variations of the stable isotopes of silicon have received little attention, which seems to be due mainly to analytical difficulties.

TILLES (1961a) extracted the silicon by fusion with sodium carbonate flux. The silicon was converted into $BaSiF_6$, which was decomposed by heating to give SiF_4, and admitted to the ion source.

Measurements by ALLENBY (1954) and REYNOLDS and VERHOOGEN (1953), using single collection of ions, differed by a factor of about 5 in the observed difference among similar kinds of samples. TILLES (1961a, b) found a total variation of $^{30}Si/^{28}Si$ ratios of about 5‰.

From Fig. 19 (after TILLES, 1961a, b) we can easily deduce that basaltic rocks are lighter in silicon isotope composition than granitic rocks.

Differences in quartz, biotite, and feldspar up to 3‰ were found in rocks of the Sierra Nevada batholith. In those rocks where the differences between the minerals were large, the observed order of enrichment, from lowest to highest ^{30}Si abundance, was biotite, quartz, and feldspar.

TILLES (1961b) studied several samples of quartz and feldspar from a zoned pegmatite. The feldspar samples were in all enriched with

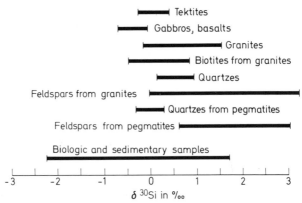

Fig. 19. Silicon isotope ratios of natural substances. (After TILLES, 1961a and b)

respect to quartz (from 0.6 to 2.7‰). The feldspar enrichment increased systematically from the outer edge of the pegmatite to the inner quartz core boundary. This finding may be explained by assuming crystal growth from a interconnected vapor phase, with silicon isotope fractionation between solid and vapor phases.

A few measurements of sedimentary and biogenic silicon showed a wide scatter of values throughout the observed range of natural variation.

A more complete understanding of the measurements requires additional data from natural and laboratory systems.

IX. Chlorine and Bromine

1. Chlorine

Chlorine has two stable isotopes with the following abundances (BOYD et al., 1955):

$$^{35}Cl: \quad 75.53\% ,$$
$$^{37}Cl: \quad 24.47\% .$$

Chlorine plays a key role in chemical atomic weight determinations. Possible differences in the natural isotope ratio may have great importance in the determination of atomic weights.

Theoretically we would expect small variations in the isotopic composition of chlorine, for instance, in the late precipitations of evaporites, analogous to sulfate-containing evaporites. HOERING and PARKER (1961), MORTON and CATANZARO (1964), and OWEN and SCHAEFER (1955), however, reported no measurable differences in the isotopic composition of chlorine.

2. Bromine

The two bromine isotopes, ^{79}Br and ^{81}Br (BAINBRIDGE and NIER, 1950) have almost equal abundances:

$$^{79}Br: \quad 50.52 ,$$
$$^{81}Br: \quad 49.48 .$$

CAMERON and LIPPERT (1955) have not found any detectable differences in the isotopic composition of bromine.

X. Alkali and Earth Alkali Elements

The results of the relatively few and sometimes contradictory isotope measurements of the alkali and earth alkali elements are combined in one chapter. We are still unable to give a well-defined picture of the possible, naturally occurring differences. This information is lacking for two main reasons: 1) the theoretically expected variations in isotope abundances should be relatively small, and 2) the reproducibility of mass-spectrometric measurements of solid substances is in most cases not better than the differences expected theoretically. It should be noted that large variations in isotope ratios arise during solid source mass spectrometry due to temperature-independent kinetic effects. The alkali elements and the alkali earth elements in the Periodic System have several chemical features in common. For stable isotope research the relatively high volatility and mobility of alkali elements and the wide occurrence of both alkali and earth alkali elements in nearly all geological environments are the most interesting similarities.

Variations in the isotopic composition of lithium, potassium, magnesium, and calcium might occur in nature. Differences in the isotope abundance of lithium in meteorites may reveal important aspects of the nucleosynthesis of lithium. Potassium appears to be mobile under the conditions in the earth's crust. The movement of potassium seems to be possible by magma, solutions, and diffusion. A variation of the natural calcium and magnesium isotope abundances caused mainly by biological and kinetic effects could be expected, which might give insight into such interesting processes as carbonate precipitation and dolomite formation.

1. Lithium

Lithium has two stable isotopes with the following abundances (BAINBRIDGE and NIER, 1950):

$$^6Li: \quad 7.52\% ,$$
$$^7Li: \quad 92.48\% .$$

The significance of measuring the $^7Li/^6Li$ ratio in meteorites and comparing it with terrestrial lithium has been recognized for many years because of the interesting nucleosynthesis of lithium.

Summarizing from the more recent papers, it may be stated that, with the more sophisticated techniques, no systematic difference has been found between meteoritic and terrestrial lithium (SHIMA and HONDA, 1963, 1966; KRANKOWSKY and MÜLLER, 1964; 1967; POSCHENRIEDER et al., 1965; DEWS, 1966; BALSIGER et al., 1968; GRADSZTAJN et al., 1968). BALSIGER et al. (1968) believe that lithium isotope variations, if they occur at all, can be of only local importance and must have a sporadic character. However, no definitely proven variations have yet been found.

2. Potassium

Potassium has three naturally occurring isotopes with the following abundances (NIER, 1950):

^{39}K: 93.08% ,
^{40}K: 0.0119% ,
^{41}K: 6.91% .

^{40}K is radioactive and decays dually, producing stable ^{40}Ca and stable ^{40}Ar.

The conclusion that the ^{39}K/^{41}K ratio is constant in nature, which has been generally accepted for a long time, may need to be revised, since several indications do not support this assumption.

Rocks originating from great depths may have lower ^{39}K/^{41}K ratios than rocks of crustal origin (LETOLLE, 1962). A similar trend on micas from genetically different rocks has been observed by LEBEDEV et al. (1966).

SCHREINER and VERBEEK (1965) described different ^{39}K/^{41}K ratios for granites and sediments. The results would be consistent with a diffusive transfer of potassium from the granite into the sediment, which would result in a preferential enrichment of the lighter isotope in the sediments. A similar relationship in the contact zone between a granite and an amphibolite has been demonstrated by VERBEEK and SCHREINER (1967), and SCHREINER and WELKE (1971).

LETOLLE (1963) argued that in the precipitation of potassium-containing evaporites a preferential enrichment of the heavier isotope occurs relative to ocean water.

3. Magnesium

Natural magnesium is composed of three isotopes (WHITE et al., 1956):

^{24}Mg: 78.8% ,
^{25}Mg: 10.15% ,
^{26}Mg: 11.06% .

The search for naturally occurring variations of the magnesium isotopes does not seem to be hopeless, if we note the wide occurrence of magnesium, for instance in carbonates, silicates, seawater, evaporites, and biological samples. However, all samples analyzed so far lie in the range of reproducibility of the mass-spectrometric measurement (SHIMA, 1964; BUCHS et al., 1965; CATANZARO and MURPHY, 1966; TAKEMATSU et al., 1967).

The discrepancy between the above results and those of DAUGHTRY et al. (1962), who found differences up to 5%, is possibly due to instrumental errors by DAUGHTRY.

4. Calcium

Calcium has six stable isotopes in the mass-range of 40 to 48 (BACKUS *et al.*, 1964):

$$^{40}Ca: \quad 96.88\% \, ,$$
$$^{42}Ca: \quad 0.655\% \, ,$$
$$^{43}Ca: \quad 0.138\% \, ,$$
$$^{44}Ca: \quad 2.12\% \, ,$$
$$^{46}Ca: \quad 0.0046\% \, ,$$
$$^{48}Ca: \quad 0.20\% \, .$$

Several authors achieved reproducibilities of around 1% in the measurement of the $^{48}Ca/^{40}Ca$ ratios (HIRT and EPSTEIN, 1964; STAHL, 1968; STAHL and WENDT, 1968; LETOLLE, 1968; CORLESS, 1968). Although a great variety of biological and geological samples were measured, no variations in calcium isotopic composition were found.

Although the stable isotope geochemistry of calcium might reveal very interesting aspects of calcium, we need more sophisticated techniques to investigate these.

C. Variations of Stable Isotope Ratios in Nature

I. Extraterrestrial Materials: Meteorites, Tektites, Lunar Materials

Meteorites may be regarded as specimens of matter from space. They supply information on the properties and composition of matter from beyond the earth. However, meteorites also have importance for geochemistry. Because direct observations of the earth's interior are not possible, we must draw conclusions in an indirect manner. One way to do this is by analogy with meteorites.

In the solar system there are probably millions of meteorites of all sizes, from the finest dust particles to hunks measuring kilometers in diameter. Our knowledge of the composition of meteorites comes from larger bodies that are observed as they fall, or from objects that are recognized as meteorites by the special characteristics distinguishing them from terrestrial rocks.

1. Meteorites

Meteorites are classified into three groups: irons, stony-irons, and stones (for an exact nomenclature, see KEIL, 1969).

Iron meteorites are classified on the basis of their structure and, to some extent, their nickel contents, into hexahedrites, octahedrites, and nickel-rich ataxites. Iron meteorites consist predominantly of metallic iron with 5.5 to 20% of alloyed nickel, small amounts of other metals, a few percent sulfides, mostly iron sulfide (troilite), some iron carbide (cohenite), and graphite. Irons form about 5% of observed meteorite falls.

Stony-iron meteorites are mainly classified into pallasites, and mesosiderites. Members of this group are characterized by appreciable nickel-iron contents.

Stone meteorites have been classified into two groups: achondrites and chondrites. Distinguishing these two groups from each other is in most cases possible on the basis of the presence of chondrules in chondrites and their absence from achondrites. Chondrules have a spherical or elliptical shape, usually a few tenths of a millimeter to a few millimeters in diameter, containing essentially the same silicate minerals as the matrix of the chondrite in which they are embedded.

There is a minor but significant group, carbonaceous chondrites, whose composition differs from that of ordinary chondrites, mainly in water- and carbon-contents. The most interesting minor constituents of carbonaceous chondrites for stable isotope research are carbonates, sulfates, sulfides, and "organic" carbon compounds.

$^{18}O/^{16}O$ data for meteorites have been published by SILVERMAN (1951), VINOGRADOV et al. (1960), REUTER et al. (1965), TAYLOR et al. (1965), and ONUMA et al. (1972a, b).

On the basis of isotopic analyses of separated minerals, TAYLOR et al. (1965) were able to divide the stony meteorites into 3 groups.

1) basaltic achondrites, hypersthene achondrites, and mesosiderites with pyroxene δ-values from 3.7 to 4.4‰;

2) high-iron group and low-iron group chondrites, enstatite chondrites, enstatite achondrites with pyroxene δ-values from 5.3 to 6.3‰;

3) carbonaceous chondrites with a highly variable oxygen isotope composition, implying that these chondrites are much less equilibrated than ordinary chondrites.

ONUMA et al. (1972a) assumed that the $\delta^{18}O$-values of coexisting minerals (plagioclase, pyroxene, olivine) from ordinary chondrites were crystallized in oxygen isotope equilibrium. They estimated the maximum temperature attained during thermal metamorphism to be $950 \pm 100°$ C.

BOATO (1954) analyzed the deuterium-to-hydrogen ratio, D/H, of the water extracted at 180° C from carbonaceous chondrites, which falls in the range of terrestrial variations. The water obtained at higher temperatures has a much greater variation range, and in four cases is outside the terrestrial range. Corrected δD-values vary between -154 and $+270$‰. A discussion on the controversial meaning of the deuterium distribution in meteorites was later published by EDWARDS (1955) and BOATO (1956).

Carbon and sulfur are common constituents of many meteorites. They can be found in a number of minerals, and they sometimes occur in various valence states in one meteorite specimen. Carbon and sulfur isotope determinations on meteorite specimens were initiated to answer the following questions:

1) What evidence exists that isotopic variations have occurred during nucleosynthesis? 2) Do large fractionations exist similar to those found on earth? 3) Is there a trend in the fractionation pattern which might indicate biological activity?

Meteoritic sulfur usually has a rather constant $\delta^{34}S$-composition (MACNAMARA and THODE, 1950; VINOGRADOV et al., 1957; AULT and KULP, 1959; THODE et al., 1961; JENSEN and NAKAI, 1962; HULSTON and THODE, 1965; MONSTER et al., 1965; KAPLAN and HULSTON, 1966).

Troilite is the most abundant sulfur compound of iron meteorites and shows $\delta^{34}S$-values from 0 to 0.6‰ relative to Canyon Diablo troilite (KAPLAN and HULSTON, 1966). Stony meteorites also contain a wide variety of sulfur compounds. MONSTER et al. (1965) and KAPLAN and HULSTON (1966) separated the various sulfur constituents. The results are summarized in Table 8.

Table 8. Distribution and isotopic composition of sulfur in the Orgueil carbonaceous chondrite. (After MONSTER et al., 1965)

Form of sulfur	Percent of sulfur by weight	$\delta^{34}S$
$MgSO_4 \times 7\,H_2O$	2.1	−1.3
Elemental sulfur	1.8	1.5
Troilite sulfur	0.8	2.6
Other forms of sulfur	0.3	-
Total sample	5.0	0.4

In contrast to most terrestrial environments, sulfate in the Orgueil meteorite is the isotopically lightest sulfur compound. MONSTER et al. (1965) suggested that kinetic isotope effects in a sulfur-water reaction may be responsible for the genesis of sulfur compounds in meteorites. There is no evidence for biological activity having occurred in meteorites, since in this case the sulfate should have been enriched in the heavy isotope.

HULSTON and THODE (1965) and KAPLAN and HULSTON (1966) have measured the $^{33}S/^{32}S$ and $^{36}S/^{32}S$ ratios in meteoritic material. Their data give strong evidence against variations due to inhomogeneities in the processes of nucleosynthesis and due to cosmic ray effects.

Carbon isotope studies on meteorites have been carried out by CRAIG (1953), BOATO (1954), CLAYTON (1963b), VINOGRADOV et al. (1967), BELSKY and KAPLAN (1970), KROUSE and MODZELESKI (1970), and SMITH and KAPLAN (1970).

BOATO (1954) determined the isotopic composition of total carbon in carbonaceous chondrites. His results are summarized in Fig. 20. BOATO argued that whatever the actual processes were that finally produced the various types of chondrites, the decrease in ^{13}C-content with decreasing water and carbon content (shown in Fig. 20) may reflect an isotopic effect during the particular process by which the volatiles were lost.

Besides the bulk carbon isotopic composition, the various carbon phases occurring in meteorites (graphite, diamond, carbides, "organic" matter, carbonates) have been separately investigated. According to VINOGRADOV et al. (1967), the $\delta^{13}C$-values of the "organic" material vary from −22.0 to −28.7‰, whereas the graphites range from −5.0 to −6.3‰. CRAIG (1953) examined the $^{13}C/^{12}C$ ratio in cohenite from the Canyon Diablo iron meteorite and found a δ-value of −17.9‰, quite different from the $^{13}C/^{12}C$ ratio for the graphite (−6.3‰) from the same meteorite.

Carbonates which rarely occur in carbonaceous chondrites, have been analyzed by CLAYTON (1963b) and KROUSE and MODZELESKI

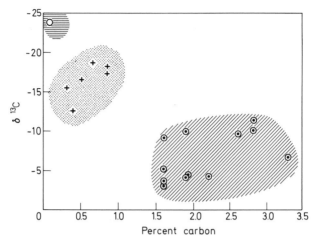

Fig. 20. Isotopic composition of total carbon in meteorites. (After BOATO, 1954 from DEGENS, 1969)
⊙ Carbonaceous chondrites Type I; + Carbonaceous chondrites Type II; ○ Ordinary chondrites.

(1970). The $\delta^{13}C$-values usually range between $+40$ and $+60‰$ relative to PDB, representing the heaviest carbon found so far on earth, whereas the isotopic composition of oxygen is not different from that found in terrestrial carbonates.

Summarizing, we may state that the bulk isotopic composition of chondrites show interesting evolutionary trends, but that the various compounds in meteorites show a very heterogenous composition. The results may imply that the carbon (or, after SMITH and KAPLAN (1970), at least that in the carbonates, the residual carbon and part of the extractable organic material) is indigenous to these meteorites and that isotopic equilibrium among the different carbon constituents has not been attained.

2. Tektites

Tektites consist of a silica-rich glass (average about 75% SiO_2). They have an unusual chemical composition, which consists of the conjunction of high silica and comparatively high alumina, potash, and lime, with low magnesia and soda. Tektites are often regarded as the product of the impact of comets or huge meteorites on the earth.

TAYLOR and EPSTEIN (1964, 1966a, 1969) have shown that the oxygen isotope composition of tektites usually ranges from 8.9 to 11.8‰. Various tektite groupings based on chemical composition and geographic occurrence all show a systematic increase in ^{18}O with decreasing SiO_2-content. These systematic correlations, as noted by TAYLOR and EP-STEIN (1969) are caused either by vapor fractionation of the tektite material during impact, already mentioned by WALTER and CLAYTON (1967) or by mixing of a SiO_2-rich igneous component and a low-SiO_2 component formed at a much lower temperature. TILLES (1961a, 1964) showed that tektites have a quite constant $^{30}Si/^{28}Si$ ratio, which is not different from the terrestrial range. FRIEDMAN (1955) determined the water and deuterium content of tektites. The D/H ratios of the water in the tektites are in the same range as that in terrestrial water.

3. Lunar Material

Stable isotope determinations on lunar material may be important for several reasons, two of which are: 1) to search for any evidence for a possible origin of the moon by fission from the earth and 2) to obtain better understanding of cosmic isotope abundances. The study of the carbon isotope composition is important to determine whether organic matter resembling that found in carbonaceous chondrites is present and to identify primordial carbon compounds that may lead to an understanding of early biochemical evolution on earth. The determination of the deuterium contents is important because a knowledge of the D/H ratios may lead to a better understanding of the development and history of the earth's hydrosphere.

EPSTEIN and TAYLOR (1970), FRIEDMAN et al. (1970), and ONUMA et al. (1970), and CLAYTON et al. (1971) analyzed the $^{18}O/^{16}O$ ratio of the silicates and oxides in the lunar material. The $\delta^{18}O$-values for the plagioclases, pyroxenes, and ilmenites are generally uniform, indicating isotopic equilibrium in the mineral assemblages. The $\delta^{18}O$-values of these minerals are comparable to terrestrial mafic and ultramafic rocks. Lunar glass, breccia, and dust are slightly enriched in ^{18}O.

The temperatures of crystallization, based on plagioclase-ilmenite temperature estimates, range from around 1070 to 1340° C.

EPSTEIN and TAYLOR (1970, 1971) also determined the $^{30}Si/^{28}Si$ ratio in lunar rocks and minerals. Plagioclase is enriched in ^{30}Si relative to coexisting pyroxene. The glass-rich dust and breccia are also richer in ^{30}Si than the whole-rock gabbros. EPSTEIN and TAYLOR suggested that this enrichment results from vapor fractionation (impact melting) of the glass during its formation.

EPSTEIN and TAYLOR (1970) and FRIEDMAN et al. (1970) determined the δD of lunar hydrogen gas, which was extremely depleted in deuterium, with δD's between -830 and 970‰. δD was hypothesized to have originated from solar wind (i.e., protons originating from the sun). The small amount of water ranged from -580 to -870‰. (FRIEDMAN et al., 1970) and -150 to -482‰ (EPSTEIN and TAYLOR, 1970). The differences between these determinations may result from different amounts of absorbed contaminating terrestrial water vapor.

EPSTEIN and TAYLOR (1970), FRIEDMAN *et al.* (1970), and KAPLAN and SMITH (1970) found unexpected high $\delta^{13}C$-values of about $+19‰$ for the lunar dust and between $+5$ and $+11‰$ for the breccia. The fine-grained rock showed a mean value of around $-24‰$, which would agree with the reduced carbon found in normal chondrites and with that found in igneous rocks.

KAPLAN and SMITH (1970) and KAPLAN and PETROWSKI (1971) analyzed the sulfur isotopic composition in lunar samples. The finegrained rock, contained the highest sulfur content, approximate average meteoritic sulfur whereas the bulk fines, lowest in S-concentration, were enriched in ^{34}S by $5‰$ and more.

II. Igneous Rocks

Magmas are the parent material of igneous rocks. They may originate through partial or complete fusion of pre-existing rocks. Besides a silicate liquid, magmas usually contain suspended crystals, and sometimes a separate gas-phase. Because of its lower density than the surrounding rocks a magma tends to rise, and differentiation (the continuous evolution of magmas from a common parent) of the melt occurs in response to the changing conditions. The most effective process of magmatic differentiation is fractional crystallization. In his classic model of igneous petrogenesis, BOWEN (1928) assigned the key role to basaltic magma. All other magmas were treated as products of fractional crystallization of the parent basaltic magma. Today, the importance of differentiation of basaltic magma by fractional crystallization to produce more siliceous and alkaline magmas is generally accepted, however, differentiation of basalt can no longer be regarded as a universal mechanism of magmatic evolution.

TAYLOR (1968a) has discussed the importance of $^{18}O/^{16}O$ measurements for igneous rock-types in relation to some of the classical problems of igneous petrology:

1) "the importance of fractional crystallization as a process of magmatic differentiation
2) whether assimilation of country rock plays an important role in magmatic evolution
3) possible differences in mode of origin of volcanic and plutonic igneous rocks
4) the effects of late-stage deuteric or hydrothermal recrystallization in igneous rocks, including the possible interaction between magmas and meteoric waters in shallow portions of the earth's crust."

First the oxygen isotope composition of the various igneous rock types is discussed in some detail (see Fig. 21). Most of the data presented in the following have been given by TAYLOR (1968a).

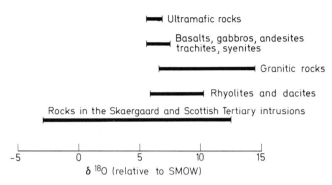

Fig. 21. Oxygen isotope composition of igneous rocks. (Mainly after TAYLOR, 1968a).

The $^{18}O/^{16}O$ ratios of *ultramafic* rocks are practically identical to those of chondritic meteorites. They exhibit a very narrow range in $\delta^{18}O$-values, from 5.4 to 6.6‰. Most of the analyzed rocks lie in an even smaller range, between 5.4 to 5.8‰. PINUS (1971) explained the small variations in the δ-values of a few olivines in different types of ultramafic rocks by the different depths of origin in the upper mantle. If this interpretation is correct in general — and not only for this limited amount of samples — then this interesting relationship would indicate some kind of zoning in the oxygen isotope composition of the mantle.

Basalts, gabbros, and *anorthosites* are isotopically almost indistinguishable from one another. Their $^{18}O/^{16}O$ ratios lie in the range from 5.5 to 7.4‰, except for some gabbroic rocks of relatively late-stage origin. The δ-value of 5.9‰ is probably the best estimate of the $\delta^{18}O$-value of primary uncontaminated basaltic magma. This value is only slightly higher than that of the average ultramafic rock, which is compatible with the hypothesis that ultramafic rocks and basaltic magmas are derived from the upper mantle, where they were in approximate isotopic equilibrium with each other at temperatures $> 1\,200°$ C.

The basaltic rocks from oceanic islands show similar features: the more SiO_2-rich and alkali-rich "differentiates" are either slightly lower or slightly higher in ^{18}O than the "parent magmas".

Andesites, trachytes, and *syenites* show $^{18}O/^{16}O$ ratios indistinguishable from basalts, indicating that these rocks may be derived from basaltic magmas by some process of magmatic differentiation. It is unlikely that any significant assimilation of sedimentary country rock could have been involved in the differentiation process.

Usually *granitic rocks* show a wider spread of δ-values (7 to 13‰) than most other igneous rocks. Plutonic granites have higher $^{18}O/^{16}O$ ratios than their volcanic equivalents because they were probably derived in large part from ^{18}O-rich sialic crust by partial melting or assimilation and/or their differentiation occurred at much lower temperatures. SHIEH and SCHWARCZ (1971) found that there are two different groups of granites in the Grenville Province of Canada: 1) higher-level intrusions into greenschist to middle amphibolite metamorphic terranes, with δ^{18}O ranging from 8.2 to 14‰, and 2) granite gneisses gradational into migmatite and upper amphibolite facies paragneiss, with δ^{18}O-values between 6.5 and 8.8‰.

MATSUHISA et al. (1972) found a similar grouping: The granites intruded into a non-metamorphic zone showed "normal" ^{18}O/^{16}O ratios between 8 and 9‰, whilst those intruded into a regional metamorphic zone are enriched in ^{18}O (δ^{18}O: 10 to 13‰). MATSUHISA et al. (1972) attributed this enrichment to isotopic exchange between the granitic magma and the surrounding metamorphic rocks during regional metamorphism. These examples demonstrate that a simple relationship among granites produced by differentiation of basic complexes with low δ^{18}O-values and granites produced by partial melting of metasedimentary rocks with high δ-values, unfortunately does not exist.

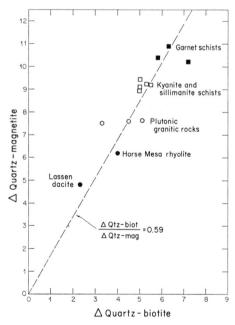

Fig. 22. Relationship between Δ-values of quartz-magnetite and quartz-biotite for quartz-biotite-magnetite assemblages from igneous and metamorphic rocks. Data are from TAYLOR and EPSTEIN (1962a), GARLICK and EPSTEIN (1967), and TAYLOR (1968a). (From TAYLOR, 1968a)

The oxygen isotope fractionations among coexisting minerals in volcanic and plutonic igneous rocks are compared in Figs. 22 and 23, using data from TAYLOR and EPSTEIN (1962b, 1963), GARLICK (1966), and TAYLOR (1968a).

At first we observe that the $^{18}O/^{16}O$ fractionations, reported in this case as Δ-values, are in general larger in metamorphic rocks than in plutonic rocks. Secondly, we note that the Δ-values are clearly smaller in rhyolite, dacite, and basalt than in granite, tonalite, and gabbro. We may generally conclude that volcanic and plutonic igneous rocks have distinctly different $^{18}O/^{16}O$ ratios. Figs. 22, 23, and 24 further indicate an approach to isotopic equilibrium in the very different mineral assemblages. However, complete isotopic equilibrium was obviously not attained in all the rocks, because otherwise all points would plot on the straight lines.

Regardless of the complicating factors, many of the Δ-values shown in Figs. 22, 23 and 24 may reflect temperatures of final equilibration or "freezing-in" of oxygen isotopes in the respective assemblages.

Table 9 shows some temperatures of "formation". If we take into account the uncertainties in the experimentally determined calibration curves and errors due to retrograde effects, especially of the feldspars and biotites, then the calculated temperature values or at least the relative differences between the values agree relatively well with other calculated temperature values based on other petrological criteria.

In several igneous rocks it is clear that coexisting quartz and feldspar and/or coexisting feldspar and pyroxene are out of isotopic equilibrium.

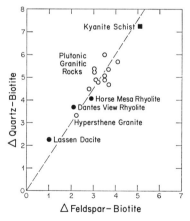

Fig. 23. Relationship between Δ-values of quartz-biotite and feldspar-biotite for quartz-feldspar-biotite assemblages in igneous rocks and a kyanite-zone schist studied by TAYLOR et al. (1963). (From TAYLOR, 1968a)

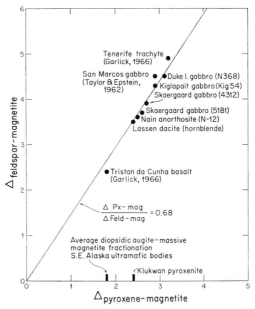

Fig. 24. Relationship between Δ-values of feldspar-magnetite and pyroxene-magnetite for feldspar-pyroxene-magnetite assemblages in igneous rocks. Data are from TAYLOR and EPSTEIN (1962b, 1963), GARLICK (1966), and TAYLOR (1968a). (From TAYLOR, 1968a)

Table 9. Oxygen isotope "temperatures" of some mineral pairs from various igneous rocks analyzed by TAYLOR and EPSTEIN (1962a), GARLICK (1966), TAYLOR (1968a) and ANDERSON et al. (1971)

Sample	Feldspar-magnetite	Quartz-magnetite	Pyroxene-magnetite	Quartz-biotite
Alkali basalt, Kilauea, Hawaii	1060			
Alkali basalt, Kilauea Iki, Hawaii	1030			
Trachyte, Tenerife	810		765	
Skaergaard gabbro	940		920	
Kiglapait gabbro	850		835	
Duke Island gabbro	770		790	
Nain anorthosite	695		950	
Lac St. Jean anorthosite	1050			
San Jose tonalite	650	665		670
Woodson Mt. granodiorite	610	670		595
Rubidoux leucogranite	615	665		895
Lassen dacite	990	1040		1305
Horse Mesa rhyolite	750	805		745
Rhyolite ignimbrite, Taupo, New Zealand	860			

Such relationships are explained if post-crystallization oxygen exchange had occurred between the mineral assemblage and a H_2O-rich fluid phase, with the feldspar undergoing exchange more easily than the coexisting quartz or pyroxene. Typical examples of oxygen-isotope exchange are the feldspars in red-rock granophyres, which are therefore not suitable for the search of oxygen-isotope changes that occur as a result of fractional crystallization in gabbro-granophyre complexes.

Oxygen-isotope variations can accompany magmatic differentiation. For example, the sequence of crystallization of minerals will influence the oxygen-isotope composition in the melt phase. It is logical that early crystallization of magnetite (a low-^{18}O mineral) should enrich the melt in ^{18}O, whereas early crystallization of quartz will deplete the melt in ^{18}O. During the last stages of crystallization $^{18}O/^{16}O$ ratios may be different from the major portion of the igneous body. Good examples for such a relationship are the Skaergaard (Greenland) and Muskox (Canada) intrusions (see Fig. 25). The average oxygen isotopic compositions of 95% of these intrusives are typical of an olivine basalt or a gabbro. While the upper parts of the Skaergaard intrusion have a lower $\delta^{18}O$-value than the major part of the gabbro, the reverse situation is found in the Muskox intrusion.

In igneous complexes a correlation generally exists between the $^{18}O/^{16}O$ ratios of SiO_2-rich differentiates and the chemical trends in volcanic

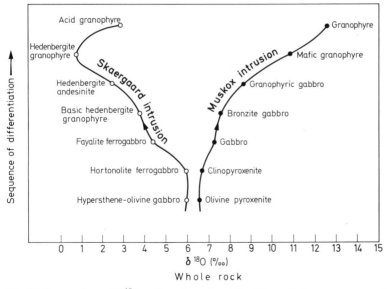

Fig. 25. Comparison of $\delta^{18}O$ values throughout the differentiation sequences of the Skaergaard intrusion, Greenland and the Muskox intrusion, Northwest Territory, Canada. (TAYLOR, 1968a)

complexes. High $^{18}O/^{16}O$ ratios accompany those trends having the lower Fe/Mg ratios, whereas ferrogabbro trends are associated with depletion in ^{18}O. Variations in oxygen fugacity may be responsible for these effects, since early precipitation of magnetite leads to both ^{18}O-enrichment and Fe-depletion in later differentiates.

Fractional crystallization may produce larger variations in the late-crystallizing minerals. Therefore, granitic pegmatites show a greater variability than do plutonic granites. For instance, quartz in pegmatites ranges from 9 to 18‰ (after TAYLOR and EPSTEIN, 1961).

Because sedimentary and metamorphic rocks have higher $\delta^{18}O$-values than igneous rocks, assimilation of country rocks into a magma chamber can affect the oxygen isotope composition of the magma. However, processes related with assimilation are very complex (e.g., dehydration and decarbonation), that is nearly impossible to predict to which extent the ^{18}O-content of a magma may be changed.

The general aspects of the $^{18}O/^{16}O$ fractionations connected with the emplacement of igneous plutonic bodies have been discussed by SHIEH and TAYLOR (1969a) and TURI and TAYLOR (1971a, b).

These investigations showed that a single pluton is not as homogenous as might perhaps be expected from the constant initial $\delta^{18}O$-value before emplacement. A marginal ^{18}O-enrichment of a pluton emplaced into ^{18}O-rich country rocks is a common phenomena. Assimilation of sialic crust is suggested as the reason why these rocks have abnormally high $\delta^{18}O$-values. On the other hand, studies by TAYLOR (1968a, 1971) and TAYLOR and FORESTER (1971) have demonstrated that some of those igneous instrusions emplaced at relatively shallow depths in the earth's crust have interacted with meteoric groundwaters during their crystallization and cooling history. Igneous rocks that have exchanged with large quantities of heated groundwaters at high temperatures can have their $^{18}O/^{16}O$ ratios lowered by as much as 10 to 15‰. The amounts of H_2O involved are estimated to be about equal in volume to that of the exchanged rock. Therefore, we may conclude that much alteration of igneous rocks is probably caused by meteoric waters rather than by H_2O released during magmatic crystallization.

The major conclusion is that there are wide variations in the manner in which granitic plutons undergo ^{18}O-interactions with their country rocks and also large variations in the magnitudes of these isotopic effects.

1. Magmatic Water

In the gas and fluid phase of magmatic melts, H_2O is the main constituent, followed by CO_2. At high pressures silicate melts can contain significant amounts of water. A water content of 5% by weight, which is tolerated by some silicate melts at water pressures of a few kilobars, corresponds to a molar concentration of approximately 30%.

Magmatic water is a solution that has thoroughly exchanged chemically and isotopically at magmatic temperatures with a magma system or an igneous rock body. The ultimate origin of the water — either juvenile, in the sense that it is coming from degassing within the lower crust or upper mantle, recycled meteoric water, or water from sedimentary rock that has subsequently exchanged both chemically and isotopically with the magma system — is not inherent in this definition.

The concept of juvenile water is a widespread one and has influenced various independent fields in igneous petrology and oregenesis. Juvenile water is commonly defined as water that has never been in contact with surface waters. Testing this concept of juvenile water with D/H in combination with $^{18}O/^{16}O$ measurements is a highly interesting topic.

One way of looking for water of possible juvenile origin is by analyzing the D/H ratios of fluid inclusions. KOKUBU et al. (1961) found δD-values of waters from liquid inclusions in basalts to be between -33 and $-60‰$. RYE (1966) reported D/H ratios of fluid inclusions in calcite, quartz, and sphalerite within a narrow range of -68 to $-83‰$.

RYE and O'NEIL (1968) analyzed the ^{18}O-content in fluid inclusions. They found that the $^{18}O/^{16}O$ ratio of the water present in oxygen-bearing minerals such as quartz and calcite indicates that the inclusions have exchanged ^{18}O with their host minerals. Only the $\delta^{18}O$-values of water in sphalerites with a very narrow $\delta^{18}O$ range between 5.8 to 6.2 seems to represent the primary isotopic composition.

Another way to search for juvenile water might be through measuring the D/H ratio of hydroxyl minerals of possible mantle or lower-crustal origin. SHEPPARD and EPSTEIN (1970) have determined the D/H ratio of unaltered phlogopites, having relatively uniform D/H ratios ($\delta D = -58 \pm 18‰$). These authors estimated the δD-value of juvenile water to be $-48 \pm 20‰$.

Based on analyses of OH-bearing igneous minerals, primary magmatic waters have been defined by TAYLOR (1967) as waters that have $\delta^{18}O$ between 7.0 and 9.5‰ and δD between -50 and $-80‰$. TAYLOR (1967) has shown that water that has isotopically equilibrated with a large reservoir of igneous silicates at magmatic temperatures has a well-defined oxygen-isotope composition between 7.0 and 9.5‰. A magmatic fluid will retain this $\delta^{18}O$-value only as long as the isotopic composition is controlled by silicate exchange at magmatic temperatures. Lower exchange temperatures (both inside and outside the igneous body) can modify this $\delta^{18}O$-value.

The hydrogen isotope composition of magmatic waters are more variable than their $^{18}O/^{16}O$ ratios. The average δD of biotite and hornblende from mafic and ultramific rocks and deep-seated batholitic rocks is typically -60 to $-90‰$ (GODFREY, 1962; TAYLOR and EPSTEIN, 1966b; SHEPPARD and EPSTEIN, 1970).

However, SHIEH and TAYLOR (1969a) and TAYLOR and EPSTEIN (1968) have shown that biotite and hornblende from shallow intrusions commonly have lower and much more variable D/H ratios (see also FRIEDMAN and SMITH, 1958; KOKUBU et al., 1961). This variation seems to be a consequence of the interaction of meteoric groundwaters of different isotopic composition with magmas or hot igneous rocks. The D/H ratios of igneous rocks and minerals are enormously more sensitive to such processes than are the $^{18}O/^{16}O$ ratios, because igneous rocks contain about 60 atom percent oxygen, and it thus requires exchange with a very large amount of H_2O to appreciably change their $^{18}O/^{16}O$ ratios.

Based on the fact that water samples have δD-values between -50 and $-80‰$, we cannot necessarily conclude that they must be of "primary magmatic origin". This is because waters of practically any origin can attain such values through isotopic exchange and other fractionation steps. In all the presently analyzed, active geothermal areas CRAIG (1963) has utilized hydrogen and oxygen isotope techniques to demonstrate the dominance of meteoric water (greater than 95%) in these aqueous systems (see also Section C, Chapter III).

2. Sulfur Isotope Composition

SMITHERINGALE and JENSEN (1963) determined $\delta^{34}S$-values of 0.1‰ for the sulfides of the normal undifferentiated basalts of tholeiitic composition. MAUGER et al. (1967) report $^{34}S/^{32}S$ ratios of intermediate rocks, alkali andesites through quartz latites, which showed a narrow spread near zero.

SCHNEIDER (1970) analyzed the sulfur-isotope composition of tholeiitic, olivine alkali, and alkali-rich basalts. In tholeiitic and alkali basalts, sulfur is present predominantly as sulfides. Alkali-rich undersaturated basalts show both sulfide and sulfate-sulfur. The $\delta^{34}S$-values of the olivine alkali basalts are concentrated closely around a mean value of 1.3‰. When the tholeiitic basalts deviate from this mean, they are enriched in ^{32}S. Due to a higher water content of the alkali-rich basalts, an oxidation of sulfur to sulfate may take place, connected with a slight enrichment of ^{34}S, if the system is open for light sulfide. SCHNEIDER concluded that the olivine alkali basalts show the least deviation from the original mantle value, which he gave as $1.3 \pm 0.5‰$.

Granites and associated sulfides show a wider variation range than the basaltic rocks. Most of the granites have small positive $\delta^{34}S$-values, while some gave abnormally low or high $^{34}S/^{32}S$ ratios (SHIMA et al., 1963; BOTH et al., 1969; SIEWERS, unpublished data).

The idea of using sulfur isotope studies to distinguish between "primary" differentiated granites and "secondary" partially melted metamorphic rocks is as old as the very first sulfur isotope measurements by THODE et al. (1949). SIEWERS (unpublished manuscript) demonstrated that this distinction cannot be done.

3. Carbon Isotope Composition

Carbonatites are magmatic carbonates (mostly calcite and dolomite) associated with alkaline rocks. It is generally agreed that they crystallized from deep-seated carbonate magmas, which is in accordance with the isotope data. The majority of carbonatites fall in the range of 6 to 10‰ for the $^{18}O/^{16}O$ ratio and -5.0 to -8.0‰ for the $^{13}C/^{12}C$ ratio (BAERTSCHI, 1957; CONWAY and TAYLOR, 1969; DEINES, 1970; KUKHARENKO and DONTSOVA, 1964; TAYLOR et al., 1967). Those samples falling outside this range may have been affected be weathering or by hydrothermal alteration.

Carbonates, often present in trace amounts in igneous rocks, show $\delta^{18}O$-values that are quite heavy and $\delta^{13}C$-values that are quite light and very variable, indicating a groundwater origin or a re-equilibration under low temperatures (O'NEIL et al., 1970; HOEFS, 1972).

HOEFS (1965) has shown that carbon occurs in at least two compounds in igneous rocks, in an oxidized form (carbonates and CO_2 in fluid inclusions) and in a reduced form (maybe elemental carbon and "organic" compounds). The reduced carbon has a strange, very light isotopic composition (-19.0 to -28.0‰, CRAIG, 1953; HOEFS, 1972). Theoretically, these δ-values can be explained by two very different mechanisms: a secondary, maybe also groundwater, origin and a primary origin of anorganically produced organic compounds. Because of the striking similarities with the carbon isotopic composition in meteorites — mainly carbonaceous chondrites — and in lunar rocks, HOEFS (1972) has favored the second possibility.

Since diamonds need high pressures for their formation, their δ-values (-3.0 to -8.0‰), like those of carbonatites, may be regarded as the primary carbon isotope ratio on earth (CRAIG, 1953; WICKMAN, 1956; VINOGRADOV et al., 1965). (See also Section C, X.)

BOTTINGA (1969b) has calculated the carbon isotope fractionation for graphite, diamond, and CO_2. In accordance with the theoretical calculations, the few measurements on diamond and graphite in contact with each other always show a slight enrichment in ^{13}C in diamond with respect to graphite (VINOGRADOV et al., 1966; VINOGRADOV et al., 1967;

and HOERING, 1961). We may, therefore, tentatively propose that diamond may form from graphite; although there are other petrological arguments, which do not favor this assumption.

III. Volcanic Gases and Geothermal Waters

Most silicate magmas solidify over a large temperature range. The residual material consists of a fluid made up predominantly of water containing substances with particularly stable molecules that are gaseous at magmatic temperatures and compounds that are very soluble in water. A fluid of this type may be produced in small amounts by the crystallization of basaltic magma and in larger amounts by the crystallization of a granitic magma.

It might be expected that the residual fluid from a crystallizing magma should show a concentration of substances of low melting points, the so-called volatiles, which were originally dissolved in the silicate melt. The general nature of these volatiles can be inferred from the composition of volcanic gases. Gases collected directly from volcanic vents or from cooling lava flows, and small quantities of trapped gas that may be obtained by heating lava specimens, all have approximately the same composition. In all instances, H_2O is the main component, CO_2 being the next most abundant. Sulfur may occur as SO_2, as elemental sulfur or as H_2S. From thermodynamic considerations it appears that in a mixture of water-sulfur-carbon in proportions found actually in nature, sulfur will be present as SO_2 at high temperatures and as H_2S at low temperatures.

One of the principal conclusions drawn from stable isotope studies of fluids in geothermal systems is that most hot spring waters (maybe 95% and more) are not primary magmatic waters but meteoric waters derived from local precipitation (CRAIG et al., 1956; CRAIG, 1966; SHEPPARD et al., 1969). Most hot spring waters have deuterium contents similar to those of local precipitations, but are usually enriched in ^{18}O by isotopic exchange with the country rock at elevated temperatures. This so-called oxygen-isotopic shift is demonstrated in Fig. 26.

However, in areas where the D/H ratios of the local meteoric waters are similar to "magmatic" water values, stable-isotope techniques cannot distinguish conclusively between the meteoric-hydrothermal and the magmatic-hydrothermal solutions.

The knowledge of the isotopically competitive processes involving the different compounds of carbon and sulfur found in volcanoes and fumaroles is of great importance in understanding carbon and sulfur isotope geochemistry, since it is by volcanism that much of the carbon and sulfur is injected into the earth's crust.

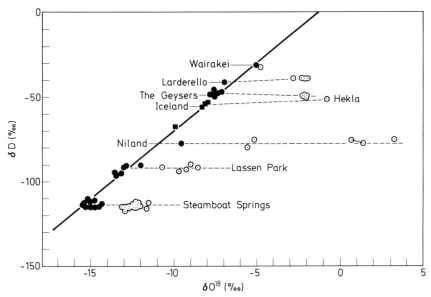

Fig. 26. Observed isotopic variations in near-neutral chloride type geothermal waters and in geothermal steam. ■: local meteoric waters or slightly heated nearsurface ground waters; ●: hot springs or geothermal water; ⊙: high temperature, high pressure, geothermal steam. Niland = Salton Sea Geothermal Area. (After CRAIG, 1963)

In volcanic emanations and probably in primary magmatic gases, CO_2 is the next largest component after H_2O. From material balance calculations (Section C, X) and from comparison with isotopic data of carbonatites and diamonds, it may be postulated that this CO_2 has a $\delta^{13}C$-value around $-7‰$. However, as may be seen from Fig. 27, the mean $\delta^{13}C$-value of CO_2 from geothermal areas varies between -3 and $-5‰$. On the other hand attempts to sample CO_2 from liquid lavas (NAUGHTON and TERADA, 1954; WASSERBURG et al., 1963) yield a wide range in $\delta^{13}C$ between -14 and $-28‰$ (see also HOEFS, 1972). The reasons for these discrepancies are not well understood. On the one hand it is possible that the CO_2 is influenced through isotopic exchange with sedimentary carbonates (CRAIG, 1963), on the other differences of physico-chemical conditions of the primary magma could also be responsible (OHMOTO, 1972).

In addition to CO_2, CH_4 and CO have been identified in fumaroles. The CH_4 content rarely exceeds 1% of the total gas content. The isotopic composition of CO_2 and CH_4 in hot springs have been determined by CRAIG (1953, 1963) and HULSTON and McCABE (1962) (see Fig. 27).

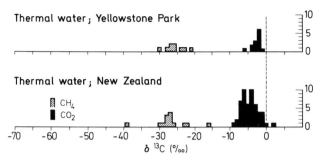

Fig. 27. Histograms of carbon isotopic composition of coexisting methane and carbon dioxide in thermal waters. (Data from CRAIG, 1953, 1963, and HULSTON and McCABE, 1962)

From Fig. 27, we may conclude that the carbon isotopic composition of geothermal methane is significantly different from those of other natural methanes (see Section C, VII). This conclusion indicates that isotopic exchange probably occurs between CO_2 and CH_4. If isotopic equilibrium is affirmed then we can calculate a temperature of the last isotopic equilibration between CO_2 and CH_4, using BOTTINGA's calculated fractionation factors (BOTTINGA, 1969a). The isotopic temperatures given by HULSTON and McCABE (1962) agreed rather well with direct temperature measurements.

The most important sulfur-containing compounds in the magmatic gas phase are H_2S and SO_2. S_2 and SO_3 are present in comparatively minor amounts. The SO_2/H_2S ratios and the fractionation factors between SO_2 and H_2S differ with varying temperatures (SAKAI, 1957; SAKAI and NAGASAWA, 1958; RAFTER et al., 1958a, b; NAKAI and JENSEN, 1967; SCHOEN and RYE, 1970). If isotope equilibrium has been reached, it is possible to calculate an "equilibrium temperature" from the calculated fractionation factors between coexisting H_2S and SO_2 (SAKAI, 1957).

In areas of recent and ancient volcanic activity, where elemental sulfur is present in appreciable amounts, somewhat different fractionation processes may occur. Magmas which erupt from great depths may bring up sulfur of deep crustal or even mantle origin to the surface. This sulfur theoretically should be only slightly fractionated and thus have a sulfur isotope composition close to the value of primordial sulfur. However, the statement of a simple genetic relationship between the parent magma and the fumarolic products simply on the ground that most $\delta^{34}S$-values for fumarolic sulfur compounds remain around zero may be in error since the processes occurring during the ascent of the gases are unknown and their influence on the isotopic composition cannot be reconstructed.

Atmospheric oxygen might have also influenced the distribution and isotopic composition of the sulfur compounds.

PUCHELT et al. (1971a) concluded that a given volcanic area appears to have a typical sulfur isotope ratio. The data of PUCHELT et al. (1971a) do not favor equilibrium conditions between the different sulfur species in the geothermal areas investigated. For instance in some pairs of sulfate and elemental sulfur the sulfate is isotopically lighter than the elemental sulfur. Kinetic effects appear to be responsible for the observed values, $H_2^{32}S$ should be oxidized faster than $H_2^{34}S$. The data of SCHOEN and RYE (1970) imply that, at least in Yellowstone Park, most of the hydrogen sulfide oxidizes to sulfur inorganically, whereas elemental sulfur oxidizes to sulfuric acid biologically.

IV. Ore Deposits

The generation of an ore deposit that involves a hydrous fluid has various aspects, some of which are: 1) A source of the ore, 2) the dissolving of the ore and other components in the fluid phase, 3) the migration of the ore-bearing fluid, and 4) the selective precipitation of the ore constituents in favorable environments.

Many geologists have tried to explain ore deposits by means of relatively simple models. However, the processes by which minor constituents of the crust become concentrated into bodies of economically valuable ore are very complex. The water, dissolved salts, metals, sulfide, and other critical constituents of each deposit may be derived from two or even more sources.

A major geological problem is to explain the formation of the relatively large accumulations of nearly insoluble sulfides. The low solubilities of many metal sulfides has raised the important question as to how they can be transported in large quantities in aqueous solutions without the need for gigantic volumes of water.

One approach to the classification of ore deposits is based mainly on δD and $\delta^{18}O$ determinations of the ore fluid. Five typical ore-generating systems have been reviewed by WHITE (1968) with respect to the origin of the water:

"1) Providencia, Mexico: close magmatic affiliations and a probable magmatic source for its ore fluid, metals, and sulfide.

2) Mississippi Valley type: connate water dominant in the ore fluid during sulfide and fluorite stages, and likely sedimentary rock sources for most critical constituents.

3) Salton Sea geothermal system, California: meteoric water dominant in the ore fluid; probably most of the dissolved metals and salts are

derived from associated sediments, but sulfur and some water and metals may be volcanic.

4) Red Sea geothermal system: seawater dominant in the ore fluid, with dissolved salts and metals from associated evaporites and clastic sediments.

5) Nonesuch Shale, Michigan: connate water and copper probably introduced diagenetically soon after the shale was deposited."

1. Sulfur Isotope Ratios of Ore Deposits

A great deal of literature on sulfur isotope ratios concerns the origin of sulfide ores. It becomes increasingly difficult to give a survey of the sulfur isotope work done on ore deposits. The following is a list of only some of the numerous papers:

ANGER et al. (1966), AULT and KULP (1959, 1960), BOTH et al. (1969), BROWN (1967), BUSCHENDORF et al. (1963), DECHOW (1960), DECHOW and JENSEN (1965), FRIEDRICH et al. (1964), GAVELIN et al. (1960), von GEHLEN et al. (1962), von GEHLEN (1965), GROSS and THODE (1965), HOEFS and GRAESER (1968), JENSEN (1957, 1958, 1959, 1967), KAJIWARA (1971), KULP et al. (1956), LUSK and CROCKET (1969), NIELSEN (1968a), PINCKNEY and RAFTER (1972), ROBERTSON and CUMMING (1968), RYZNAR et al. (1967), SANGSTER (1968), SASAKI and KROUSE (1969), STANTON and RAFTER (1966), TATSUMI (1965), TUPPER (1960), WANLESS et al. (1960), VINOGRADOV et al. (1956), WHITE (1968).

Many of these investigations have been undertaken to discover systematic isotopic variations between ores of various types.

The goal of many determinations was to use $^{34}S/^{32}S$ ratios in distinguishing between high temperature magmatic or magmatic hydrothermal and low temperature biogenic deposits. Another aim was the use of sulfur isotopes to demonstrate the syngenetic or epigenetic origin of certain ore deposits. Some investigators have tried to find criteria for equilibrium, temperature of formation, and other related problems. On the whole there seems to be too much overlap between the different classes of orebody for sulfur isotope ratios to be regarded as unambiguous indications of origin, though they may limit the range of possibilities.

On the basis of $^{34}S/^{32}S$ determinations some vague generalizations can be stated:

1) Sulfides of biogenic (sedimentary) origin exhibit two differences from the "magmatic" hydrothermal deposits, a much greater spread in $\delta^{34}S$-values, and a preferential enrichment in the lighter isotope (there are, however, "heavy" sedimentary sulfides too).

2) Sulfides probably connected with magmatism rarely vary in $\delta^{34}S$-composition by more than 5‰, even throughout a whole mining district. These deposits are probably derived from homogeneous sources, such as hydrothermal ore solutions, which seem to be almost uncontaminated by surrounding rocks.

3) Sulfides collected from two ore deposits, which according to geological evidence have the same origin, may show quite different ratios. On the other hand, certain ore provinces seem to have a typical $\delta^{34}S$-composition (for instance almost all Cordilleran hydrothermal ore deposits, e.g., porphyry coppers exhibit $\delta^{34}S$-values near zero, JENSEN, 1967).

4) Systematic regional and depth variations have been found in detailed studies of particular deposits (African Copperbelt, DECHOW and JENSEN, 1965; Quemont Mine, RYZNAR et al., 1967; Mississippi Valley, BROWN, 1967, and others). However, no convincing interpretations about the spatial distribution of sulfur isotopes have been given.

5) In a given mineralization there is a tendency for the low-temperature sulfides to have a wider range of variation as compared with the high-temperature sulfides. During progressive metamorphism there seems to take place a homogenization of the sulfur isotopes.

6) In many suites of coexisting sulfide minerals galena is the lightest, sphalerite is intermediate, and pyrite is the heaviest in sulfur isotope composition. The magnitude of the differences (normally 2 to 5‰) decreases with increasing metamorphic grade of the host rock, indicating that the minerals are in isotope equilibrium.

7) It has been tried to correlate variations of lead and sulfur isotope ratios in galenas (e.g. BROWN, 1967; ROBERTSON and CUMMING, 1968; LANCELOT et al., 1971; GREIG et al., 1971). In general, the light sulfur ^{32}S is associated with lead relatively enriched in the radiogenic isotopes 206, 207, and 208.

In these generalizations it is assumed that the ore minerals accurately reflect the isotopic compositions of the ore-forming fluids. This assumption was questioned by OHMOTO (1972), who demonstrated that the isotopic composition of sulfide minerals may be influenced by 1) isotopic composition of total sulfur in the fluids, 2) temperature, 3) fugacity of oxygen, and 4) pH of the fluids. At 250° C for example, OHMOTO (1972) could show that within geologically important f_{O_2}-pH regions an increase in f_{O_2}-value by 1 log unit or in pH by 1 unit can cause a decrease in $\delta^{34}S$-values of sulfides by as much as 20‰. However, as OHMOTO (1972) pointed out, these considerations are based on the assumptions that both chemical and isotopic equilibria are established among aqueous sulfur species and between aqueous sulfur species and precipitating sulfides. Both of these assumptions may be not valid in many geologic processes.

We shall discuss two types of ore deposits in more detail, the liquid-magmatic and the stratiform deposits.

The liquid-magmatic ore deposits, representing the most simple case in which the origin of the sulfur is in the lower crust or upper mantle, show within single deposits and between the various deposits a rather uniform distribution around zero (THODE et al., 1962; SMITHERINGALE and JENSEN, 1963; SHIMA et al., 1963). For example, THODE et al. (1962) found that sulfides in Stillwater sills have δ^{34}S-values around zero. For the Insizwa sill, South Africa (THODE et al., 1962) and for the Muskox intrusion, Canada (SASAKI, 1969) a rather large variation range has been reported.

Another important category of sulfide deposits of world-wide distribution is that of the "stratiform" deposits. Depending on the types of stratified rocks, stratiform deposits can be divided into "normal" and "volcanic" types. The normal-type deposits occur in more or less nonvolcanic marine sediments, and the volcanic-type deposits are characteristically associated with various types of volcanic rocks. While the sedimentary rocks associated with the normal type appear to have formed in relatively shallow basins on continental basements, the volcanic rocks containing the volcanic type seem to be deposited in mobile geosynclinal depressions.

A majority of the stratiform deposits studied by SANGSTER (1968) have a relatively narrow spread in δ^{34}S-values, and all ore bodies are depleted in ^{34}S relative to contemporaneous seawater sulfate.

SANGSTER (1968) suggested that all of the sulfur in "sedimentary" type strata bound sulfide deposits and probably most of the sulfur in volcanic strata bound deposits was derived from sulfate in ancient seas and was reduced to sulfide by bacterial action.

SASAKI (1970) argued that the parallel variation trend observed between sulfate and sulfide may be an indication of isotopic exchange reaction between seawater sulfate and ore sulfides. Since the amount of sulfur from ocean water would normally be much larger than that from volcanic sources, it is expected that the isotopic compositions of stratiform ore deposits are largely controlled by the δ^{34}S of seawater.

2. Hydrothermal Carbonates

Carbonates, mainly calcite, are some of the most common gangue minerals. The stability field of carbonates is large, even at very low CO_2 fugacities and high temperatures.

If we assume that an isotopic equilibrium between hydrothermal carbonates and fluids was established, then the carbon and oxygen isotopic composition of the carbonates associated with ore minerals may be used to obtain more detailed information about 1) the temperature of these carbonates as compared, for instance, with fluid inclusion studies, 2) the isotopic composition of the hydrothermal fluids involved as an indicator of their origin, and 3) the origin of the CO_2 or HCO_3^- that determined the carbon composition of these carbonates.

OHMOTO (1972) has demonstrated that the carbon isotope composition of hydrothermal carbonates should be also affected by the physico-chemical conditions of hydrothermal fluids. For instance an increase in f_{O_2} by 1 log unit or in pH by 2 units can cause a decrease of about 30‰ in $\delta^{13}C$-values of carbon-bearing minerals.

Hydrothermal carbonates have been analyzed by ENGEL et al. (1958), FRIEDMAN and HALL (1963), GARLICK and EPSTEIN (1966), RYE (1966), HALL and FRIEDMAN (1969), FRITZ (1969), and PINCKNEY and RYE (1972). The observations made by these authors may be interpreted as the result of exchange of carbonate rocks with hydrothermal fluids. Isotopic exchange of carbonates with water at elevated temperatures leads to a decrease in ^{18}O of the carbonates, the decrease being greater the higher the temperatures.

There are some indications that the $\delta^{18}O$-value of carbonates of host rocks of ore deposits seem to decrease nearer to the ore deposit. In some cases, it was suggested that this relationship might be a useful guide to ore. However, in most cases, the analyzed $\delta^{18}O$-values (and to a smaller extent the $\delta^{13}C$-values) do not follow this simple relationship (see also PINCKNEY and RYE, 1972).

High-temperature hydrothermal carbonates tend to have lower $\delta^{13}C$-values than low-temperature carbonates. A $\delta^{13}C$-value of $-7‰$ has been obtained by RYE and O'NEIL (1968) during their investigation of the Providencia (Mexico) mine for the CO_2 in inclusions in sphalerite which is consistent with a primary magmatic origin.

3. Wall Rock Alteration

The environment of ore deposition can be interpreted in part from the assemblages of alteration minerals. Wall rock alteration includes those mineralogical and chemical changes brought about by circulating solutions within the host rock of ore bodies.

Applying stable isotope techniques, the following problems are especially interesting: 1) the delineation of hypogene and supergene alteration products ("hypogene" refers to ore deposits or associated gangue and alteration products that formed in the presence of generally ascending hydrothermal solutions; "supergene" processes refer to the formation of minerals and ores near the surface by low-temperature, descending waters); 2) the relative importance of meteoric H_2O versus "magmatic" H_2O in the hydrothermal fluids; and 3) the temperatures of formation of the alteration mineral assemblages.

SHEPPARD et al. (1969, 1971) determined the D/H and $^{18}O/^{16}O$ ratios in alteration assemblages from porphyry copper deposits and other hydrothermal mineral deposits.

The following conclusions can be drawn from these studies: The D/H and $^{18}O/^{16}O$ ratios of hypogene clays show a clear-cut geographic correlation with the isotopic variations displayed by present-day meteoric waters. In many situations where the D/H values of meteoric and primary magmatic waters are significantly different, the isotopic data require that the hypogene hydrothermal solutions responsible for intermediate and advanced argillic alteration contain a large amount of meteoric water. Qualitative temperatures, depending on assumptions, have been calculated in the range of 360 to 280° C for sericitization and 460 to 390° C for potassic alteration. ESLINGER and SAVIN (1971) empirically calibrated a quartz-illite geothermometer on hydrothermally altered rocks. They obtained temperatures between about 200 and 300° C.

V. Water Cycle

One of the most interesting topics of modern geochemistry is the water cycle in nature. The importance of the role of water in nearly all geological processes cannot be overestimated.

Types of water can be classified in a number of ways. Most commonly they are grouped according to 1) origin, in terms of meteoric, connate, or juvenile water; 2) chemistry, e.g. bicarbonate, sulfate, or chloride water; and 3) total salinity, e.g., freshwater and brine.

The term "meteoric" applies to water that has gone through the meteorological cycle, which is evaporation, condensation, and finally precipitation. All continental surface waters — such as rivers, lakes, and glaciers — fall into this general category. Because meteoric water may seep into the underlying rock strata, it will also be found at various depths of the lithosphere.

The ocean, although it continuously receives the continental run-off of meteoric waters as well as rain, is by general agreement not regarded as being of meteoric nature.

The terms "freshwater" and "brine" are frequently employed. The upper boundary for freshwater is 1% salinity, which is about the maximum salt concentration tolerated by the natural freshwater fauna and flora. Waters beyond the concentration of ocean water are known as brines.

The total concentration of dissolved solids in seawaters has a mean value close to 3.5%, but concentration reaches as much as 4% in partially enclosed seas in areas of high evaporation. Most of the dissolved material is in the form of simple ions or cation-anion complexes. The principal cations, in order of abundance, are Na^+, Mg^{++} Ca^{++}, and K^+; the principal anions are Cl^-, SO_4^{--}, HCO_3^-.

As is discussed in the following, the different water reservoirs found in and on earth differ isotopically. First, however, we list some definitions concerning water of different origin (after WHITE, 1957a, b).

Meteoric water: The term "meteoric" applies to water that was recently involved in atmospheric circulation.

Connate water: Connate water is "fossil" water, which has been trapped in the sediments at the time of burial.

Formation water: Formation water is present in rocks immediately before drilling (it may be a useful nongenetic term for waters of unknown age and origin but should not replace the genetic concept of connate water).

Metamorphic water: Metamorphic water is water that is or has been associated with rocks during their metamorphism.

Hydrothermal water: This term refers to any water that is warm or hot relative to its surrounding environment and has no genetic implications.

Magmatic water: Magmatic water is in or is derived from magma. Most (or perhaps all) magmatic waters may actually be recycled through remelting or partial melting of sedimentary and volcanic rocks.

Juvenile water: Juvenile water is new water that is in or is derived from primary magma or other matter and has not previously been a part of the hydrosphere.

One can look upon the hydrologic cycle as a sequential, dynamic system in which water moves as a result of such processes as condensation, precipitation, evaporation, infiltration, and runoff. In order to understand the variations in isotopic composition in natural waters it is necessary to establish the nature of the variations in precipitations, the melting of polar ice, continental runoff, etc., which have been extensively discussed by FRIEDMAN *et al.* (1964).

The natural water cycle has been compared with a multiple-stage distillation column with reflux of the condensate to the reservoir (EPSTEIN and MAYEDA, 1953). The oceans correspond to the reservoir, and the ice fields at the Poles correspond to the highest stages of the column. (The isotopic composition of snow near the South Pole is deficient in ^{18}O by 60‰ compared with ocean water (ALDAZ and DEUTSCH, 1967).

As has been mentioned earlier, the most efficient process in creating variations in the isotope ratios of water is by vapor pressure differences.

In all processes concerning the water cycle the hydrogen isotopes are fractionated in proportion to the oxygen isotopes, because a similar difference in vapor pressures exists between H_2O and HDO on one side and $H_2^{16}O$ and $H_2^{18}O$ on the other side. Therefore, the hydrogen and oxygen isotope distributions are correlated in meteoric waters. CRAIG (1961a) has given the following relationship:

$$\delta D = 8\delta^{18}O + 10 .$$

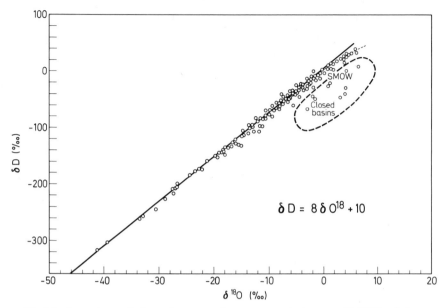

Fig. 28. Isotopic variations in meteoric waters, relative to SMOW. (After
CRAIG, 1961 a)

Most deviations of meteoric water from the normal straight line lie
on the right-hand side of the line showing low δD and/or high $\delta^{18}O$
(CRAIG, 1961 a; CLAYTON et al., 1966).

A linear isotope relationship between deuterium and ^{18}O concentration does not exist when the water is formed under non-equilibrium
conditions. For example, in a dry climate (desert), a reevaporation of
falling raindrops may increase the fractionation factors. Any process of
free evaporation is governed by kinetic factors (CRAIG et al., 1963; EH-
HALT et al., 1963). This will cause certain deviations from the linear δD-
$\delta^{18}O$ pattern normally produced by Rayleigh condensation in clouds at
liquid-vapor equilibrium.

Interesting results have been obtained by WOODCOCK and FRIED-
MAN (1963), who observed an inverse relationship between raindrop size
and deuterium content (the small raindrops are often enriched in deuterium). FRIEDMAN et al. (1962) conducted laboratory experiments to determine the length of time necessary for single water drops to reach isotopic
equilibrium with water vapor. They found a relatively rapid exchange of
deuterium between droplet and environment.

The comparison of the D/H and $^{18}O/^{16}O$ ratios shows that atmospheric precipitation normally follows a Rayleigh process at liquid-va-

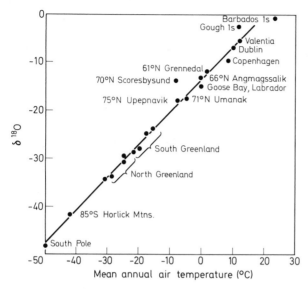

Fig. 29a. Observed $\delta^{18}O$ concentration in average annual precipitation as a function of mean annual air temperature. (After DANSGAARD, 1964)

por equilibrium. The atmospheric Rayleigh process also explains why, at higher altitudes and latitudes, fresh waters become progressively lighter isotopically, whereas tropical samples show very small depletions relative to ocean water (see Fig. 29a).

Although the isotopic composition of precipitation at any location depends on various factors, temperature seems to be the most important parameter. DANSGAARD (1964) has shown that for marine climates at medium to high latitudes the $^{18}O/^{16}O$ composition of rain averaged over a whole year varies linearly with the mean annual temperature difference between the place of evaporation (the tropical ocean) and the place of deposition. The $^{18}O/^{16}O$ composition of averaged rain decreases by 0.7‰ as the temperature difference decreases by 1° C.

Seasonal variations of the isotopic composition of rain and snow are also caused by temperature differences. The $\delta^{18}O$-values of rain and snow at a given locality show a minimum during winter and a maximum during summer. A good example of the seasonal dependence has been given by DEUTSCH et al. (1966) on an Austrian glacier, where the mean δD oscillation between winter and summer snow is 140‰.

Hydrogen and oxygen isotope determinations have provided stimulating and fruitful results in the field of glaciology (EPSTEIN and SHARP, 1959; EPSTEIN et al., 1965, 1970; GONFIANTINI, 1965; DANSGAARD et al.,

1969; PICCIOTTO et al., 1960; DANSGAARD and TAUBER, 1969; ALDAZ and DEUTSCH, 1967; LORIUS et al., 1969 and others). The measurement of D and ^{18}O is commonly used to distinguish the annual layers of ice. DANSGAARD et al. (1969) have used the ^{18}O-concentration in a Greenland ice sheet to deduce the climatic changes during the past 100000 years. The mean isotopic composition of a section of an ice core reflects the average climatic conditions existing at the surface during the period of time when the original deposit was formed (increasing δ^{18}O along the core corresponds to warming climate and decreasing δ^{18}O corresponds to cooling climate). GOW and EPSTEIN (1972) demonstrated that different ice types (e.g. glacial ice and fresh-water ice formed from desalinated sea water) can be distinguished by stable isotope analysis.

Fractionations due to freezing-point differences play a minor role in producing variations on a geographic scale, because the differences produced are not so great compared with differences due to vapor pressure differences, and because ice is relatively immobile and usually melts not far from the place in which it forms.

FRIEDMAN et al. (1964) discussed changes in the isotope concentration of water undergoing any change of state. The fractionation factor for deuterium between ice and water has been estimated and determined experimentally by several authors (e.g., O'NEIL, 1968; ARNASON, 1969). From these results we may deduce that ice will contain 2% more deuterium than the water from which it is freezing.

1. Ocean Water

Ocean water exhibits a very narrow range in isotopic composition, less than 1% for δD and 1‰ for δ^{18}O. However, evaporation processes strongly affect the isotope composition, since they cause a preferential depletion in the lighter isotopes, which become enriched in the vapor phase. Consequently, the remaining water will be heavier, i.e., highly saline waters generally have the greatest D and ^{18}O content. Low salinities, which are caused by a dilution of fresh waters, correlate with low D and ^{18}O concentrations (EPSTEIN and MAYEDA, 1953; REDFIELD and FRIEDMAN, 1965). There is one interesting feature that is difficult to explain: Deep-water masses originating in the Northern hemisphere, definitely contain a slightly larger amount of deuterium than those of Antarctic origin (REDFIELD and FRIEDMAN, 1965). All factors governing isotopic variations in ocean water have been discussed extensively by CRAIG and GORDON (1965).

Secular changes in the isotopic composition of the oceans have occurred as a result of the variable extent of continental glaciation. The maximum effect during the Pleistocene age has been estimated by various authors to be 0.5 to 1.5‰, in ^{18}O. Going back from Recent to Paleozoic time it seems rather certain that ocean water, within small variations, had a constant isotopic composition. Greater variations of several per mil and more may have been possible in Precambrian time. (PERRY, 1967, suggested that the low ^{18}O-contents of Precambrian massive cherts reflect lower ^{18}O-contents in the Precambrian oceans.) Assuming a constant rate of weathering and sedimentation throughout the earth's history, SAVIN and EPSTEIN (1970b) concluded from material balance computations that sedimentary processes have probably caused a depletion of about 3‰ in ^{18}O in the world's oceans. (As will be discussed under "Sedimentary Rocks", sediments, especially their authigenic components, are appreciably heavier in ^{18}O than weathered igneous rocks.) SAVIN and EPSTEIN (1970b) argued that most of the depletion of ^{18}O took place during Precambrian time, and only about 0.6‰ since the end of the Precambrian.

Since connate waters have been defined as ancient ocean water trapped in the sediments, they should have an isotopic composition similar to ocean water. However, the processes involved in the development of saline formation waters (brines) are complicated by the extensive chemical changes that have taken place in the brines after sediment deposition. CLAYTON et al. (1966) concluded from the relationships between δD and $\delta^{18}O$-variations and chemical compositions of the brines that the deuterium content has not been greatly altered by exchange or fractionation processes but that extensive oxygen exchange has taken place between water and reservoir rocks.

CLAYTON et al. (1966) demonstrated that oil field saline formation waters are meteoric and not connate waters. However, the problem of the origin of oil field brines still seems problematical. Many natural brines, originally interpreted as connate waters, may be meteoric waters that have undergone isotopic exchange and enrichments in salts.

2. The Isotopic Composition of Dissolved Compounds in Natural Water

a) Carbon Species in Water

In addition to organic carbon, four other carbon species exist in natural water: dissolved CO_2, H_2CO_3, HCO_3^-, and CO_3^{--}, all of which tend to equilibrate with each other. As has already been mentioned, the concentration and the isotopic composition of the single compounds vary with temperature and pH.

HCO_3^- is the dominant species in ocean water. Assuming that pH = 8.2, and $T = 25°$ C, the isotopic composition of ocean water in equilibrium with atmospheric CO_2 should be around zero relative to

PDB. DEUSER and HUNT (1969) report δ^{13}C-values of oceanic inorganic carbon between -1 and $+2‰$ and find a correlation below 200 meters water depth between the concentration of dissolved oxygen and the δ^{13}C, which is due to the oxidation of organic matter.

Fresh waters show a δ^{13}C-range for dissolved inorganic carbon of -5 to $-11‰$ (SACKETT and MOORE, 1966). Brackish waters of estuaries increase in ^{13}C with salinity. Therefore, carbon isotope composition may be used to characterize mixing of various water masses. A variable portion of bicarbonate in fresh waters is derived from biogenic sources in the soil (VOGEL, 1959 and MÜNNICH and VOGEL, 1962). Locally, ground waters may be affected by dissolution of "heavy" limestones.

b) Sulfate in Ocean Water and Fresh Water

Sulfur Isotope Composition. Present ocean water with its large sulfate reservoir has a fairly constant isotopic sulfur composition of $+20‰$. An interesting question is whether an isotope fractionation occurs in the sulfate during evaporation of seawater. NIELSEN and RICKE (1964) showed that late evaporites within the different evaporation cycles are depleted in ^{34}S by about $2‰$ relative to the normal stages. NIELSEN (data to be published), through experimental data, and THODE and MONSTER (1965), through calculated data, supported this relationship. Nevertheless, the differences that might occur in the late stages may be neglected if we consider the gypsum precipitate — brine relationship. If we assume that evaporite sulfates preserve the δ^{34}S-values of the ancient oceans — and there are reasonable arguments to do so — then we may conclude that gypsum, anhydrite, and other sulfate-containing evaporite minerals give us information about the isotopic composition of oceanic sulfate during the geologic past.

The surprising fact is that the isotopic composition of oceanic sulfate has not remained constant during the geologic past. The general trends of evaporite δ^{34}S-evolution are (after NIELSEN, 1965 and THODE and MONSTER, 1964): high δ-values ($+20$ to $+30‰$) in the early Paleozoic, decreasing to $+11‰$ in Permian time, rapidly increasing in the early Mesozoic, and later on slight oscillations around the present value of $+20‰$. Figure 29 b shows a more detailed picture.

The mechanisms controlling the decrease in Permian time and the abrupt increase at the Paleozoic-Mesozoic boundary are not understood. Some speculations are discussed at the end of this chapter.

The trend of δ-values of sulfate evolution in the world's oceans is so consistent, especially the very narrow range of Permian δ-values, that it has been used successfully to determine the age of unknown salt deposits and the origin of sulfate-containing formation water (MÜLLER *et al.*, 1966, NIELSEN, 1968 c).

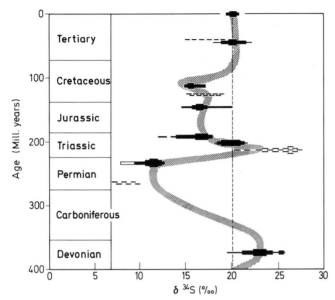

Fig. 29b. $\delta^{34}S$ age curve of oceanic sulfate. The black areas represent values of evaporite sulfates mainly from Central Europe. Shaded areas represent values obtained from evaporite basins of presumably not completely marine origin. (After NIELSEN, 1968c)

NIELSEN (1968c, 1972b) has described some practical applications in Germany, where the conditions for such "dating" are favorable, mainly for two reasons: 1) Due to the paleogeographical development the number of possible evaporite ages is limited and covers those formations that exhibit extreme δ-values (Fig. 30). 2) In many cases the question is only to distinguish between mobilized Zechstein and nonmobilized younger evaporite materials. Due to the extreme δ-value of Zechstein sulfate this question may be solved relatively easily.

In most fresh water the sulfate ion is the second or third most abundant anion, being exceeded only by HCO_3^- and in some cases by silicate.

Nonmarine waters and nonmarine evaporites differ widely in their absolute content and their $^{34}S/^{32}S$ ratios, depending on the wide variety of sulfur sources of the drainage area feeding the rivers or lakes (HOLSER and KAPLAN, 1966).

Oxygen Isotope Composition. Measurements of large numbers of samples of oceanic sulfate indicate a very constant oxygen isotopic composition, averaging around 9.7‰ (RAFTER and MIZUTANI, 1967; LONGINELLI and CRAIG, 1967; LLOYD, 1967). LONGINELLI and CRAIG (1967) have suggested that the uniform isotopic composition of oceanic sulfate may represent long-term isotopic equilibrium between oceanic sulfate

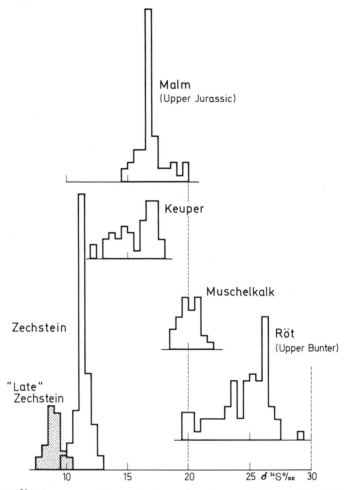

Fig. 30. δ^{34}S histograms of evaporite formations in central Europe. The Zechstein histogram is strongly reduced in its vertical scale. The shaded histogram refers to sulfates occurring in the K-Mg facies of the Zechstein evaporation cycles. (After NIELSEN, 1968 c)

and ocean water. However, if isotopic equilibrium between water and sulfate controlled the isotopic composition of sulfate, the latter should have either a δ-value of $+24‰$ deduced from theoretical calculations (UREY, 1947) or a δ-value of $+38‰$ deduced from experimental determinations (LLOYD, 1967). LLOYD (1967, 1968) proposed a model in which the bacterial turnover of sulfate in the sulfur cycle plays a decisive role in oxygen isotope composition of sulfate. RÖSLER et al. (1968) and PILOT et *al.* (1972) have shown that the oxygen isotope composition of ancient evaporite sulfates is not constant. However, more data are needed before an "age-effect" may be established.

Sulfates from nonmarine environments range over a wide spectrum. LONGINELLI and CORTECCI (1970) determined the $\delta^{18}O$ of the dissolved sulfate in river waters. Large isotopic variations in both oxygen and sulfur take place along the course of the analyzed rivers in the direction of a progressive enrichment in the heavy isotopes. Several factors seem to be responsible for these variation, i.e., human and microbiological activity.

c) Evolution of the $\delta^{34}S$ of Ocean Water Sulfate during the Geologic Past

As has already been mentioned, the δ-value of the ocean sulfate has not been constant during the earth's history. The total range of variation has been at least 20‰. HOLSER and KAPLAN (1966) have stated that the possible implications of any such changes reach far beyond the significance of the change in sulfate itself and have suggested some implications.

NIELSEN (1965) has postulated that the normal condition of the ocean is one where the sulfur input rate is equal to the output rate and where $\delta(\text{input}) = \delta(\text{output})$. In order to change $\delta(\text{ocean})$ it is necessary to increase the output rate of bacterial sulfate reduction and/or the input rate of sulfur to the ocean reservoir. A similar model has been given by HOLSER and KAPLAN (1966). According to these authors, changes of $\delta(\text{ocean})$ reflect changes in the total sulfate content of the ocean. These authors postulated that between Cambrian and Permian times, this content should have increased by nearly a factor of 2 and that between Permian and Recent time it should have decreased by some 30‰.

REES (1970) developed a somewhat different model in which the process of evaporite formation plays a key role. With two mechanisms competing for the removal of sulfate from ocean water, one of which involves isotope fractionation (sulfide formation) and one of which does not (evaporite formation), the net fractionation produced in the ocean water reservoir will depend on the relative extent of these mechanisms.

VI. Atmosphere

The atmosphere consists of a mixture of a number of gases. The chemical composition is quite simple, being made up almost entirely of three elements: nitrogen, oxygen, and argon. Other elements and compounds are present in amounts that, although small, are nevertheless significant in terms of the geochemistry of the atmosphere. The atmosphere is moderately homogenous in composition, except for water, which varies with location and season as well as elevation. The average abundance of some important constituents of the atmosphere is shown in Table 10.

Table 10. Average of some important constituents in the atmosphere (on a water-free basis)

Gas	Abundance by volume in %
N_2	78.09
O_2	20.95
Ar	0.93
CO_2	0.03
H_2	0.00005
N_2O	0.00005

There seems to be little question now that the constituents of the atmosphere have been largely degassed from the earth's mantle. From observations of volcanic and of cosmic gases, different concepts about the chemical composition of the primeval atmosphere agree that free oxygen was absent and that reducing conditions prevailed. The first free oxygen was probably produced through photochemical dissociation of water vapor in the upper atmosphere, which would produce oxygen and hydrogen, with the hydrogen escaping into outer space. However, the free oxygen probably did not accumulate in the atmosphere, but was used up in oxidizing the more-reduced constituents of the atmosphere. This stage came to an end when oxygen production exceeded oxygen use, which probably happened when photosynthesis started to produce free oxygen on a world-wide scale.

The amount of argon, which is 99.6% ^{40}Ar, in the atmosphere is anomalously high when compared with that of the other inert gases. This is evidently due to the production of ^{40}Ar by the radioactive decay of ^{40}K throughout geological time.

Atmospheric nitrogen collected from many altitudes shows a constant isotopic composition (DOLE et al., 1954). Air samples collected over a six-month period at several locations had a constant ratio, within 0.2‰ (HOERING, 1956).

HOERING (1957) determined the isotopic composition of the ammonia and nitrate ions in rain. He demonstrated that the ^{15}N-content of the two nitrogen species varies from rain to rain and that there was no apparent relationship between the isotopic composition of nitrogen in the nitrate ion and in atmospheric nitrogen. He observed a constant kinetic isotope effect of about 1.004 between the ammonia and the nitrate ion.

Atmospheric oxygen has a rather constant isotopic composition with a $\delta^{18}O$-value of $+23.5$ (DOLE et al., 1954; KROOPNICK and CRAIG, 1972). Oxygen dissolved in ocean water is further enriched in ^{18}O by nearly 1‰ at 0° C (KROOPNICK and CRAIG, 1972). UREY (1947) calculated the equilibrium constant for the exchange reaction between atmospheric oxygen and water to be very close to unity. This means that atmospheric oxygen cannot be in equilibrium with the hydrosphere. Therefore, the enrichment of ^{18}O in free O_2, the so-called Dole effect, must have another explanation.

Oxygen isotope studies proved that photosynthetic oxygen originates from water, not from CO_2:

$$6CO_2 + 6H_2O^* \underset{\text{respiration}}{\overset{\text{photosynthesis}}{\rightleftharpoons}} C_6H_{12}O_6^* + 6O_2 \,.$$

It is well-known that photosynthesis is the main source of atmospheric oxygen. The intensity of this process is high enough to ensure a complete exchange of atmospheric oxygen in 3 to 4 thousand years. RABINOWITCH (1945) suggested that the high ^{18}O-content of atmospheric oxygen was due to preferential consumption of ^{16}O during respiration of plants and animals which, at a steady state, must remove oxygen equal in composition to photosynthetic oxygen to the atmosphere. In other words, the ratios for photosynthetic oxygen delivered to the atmosphere and the oxygen extracted from the atmosphere are equal.

The investigations of LANE and DOLE (1956), who found that the fractionations accompanying respiration are organismdependent, are compatible with this hypothesis.

If this explanation is correct, then there might be some fluctuations in the absolute concentration and in the ^{18}O-content of the atmosphere from Cambrian to Recent time, in accordance with varying rates of photosynthesis during the earth's history (WELTE, 1970).

The CO_2 content of the atmosphere controls many processes of geological importance (e.g., pH of ocean water and the "greenhouse effect" of shielding reflected sun energy) althoug it is only present as 0.03%.

Daily, seasonal, secular, local, and regional changes in atmospheric CO_2 content have been observed as regular fluctuations. Careful determinations by KEELING (1958, 1960, 1961) showed that daily variations, depending on respiration, exist over continents. Respiration of plants reaches a distinct maximum around midnight or in the early morning hours. Respiratory plant CO_2 has a $\delta^{13}C$-value between -21 and -26‰.

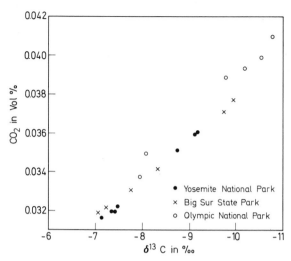

Fig. 31. Relation between carbon isotope ratio and concentration of atmospheric CO_2. (After KEELING, 1958)

At night, when respiration of plants reaches the maximum values, we should find a measurable contribution of respiratory CO_2 in near-surface regions. This relation between CO_2 content and $\delta^{13}C$ is demonstrated in Fig. 31.

As we can see from Fig. 31, at the minimum concentration of around 314 ppm, atmospheric CO_2 has a $\delta^{13}C$-value of around -7, which is identical to marine air.

FRIEDMAN and IRSA (1967) have found significant differences in the absolute concentration and the isotopic composition of CO_2 sampled at street level in New York City. However, when rapid mixing and diffusion of gaseous combustion products is possible, there seems to be little variation in the $^{13}C/^{12}C$ ratios of the CO_2.

Atmospheric CO_2 is the compound containing the heaviest oxygen, with a δ-value of $+41‰$, which means that atmospheric CO_2 is in approximate equilibrium with ocean water at $25°$ C. Variations of about $1‰$ have been observed by KEELING (1961), but until corrections for possible N_2O contents (CRAIG and KEELING, 1963) have been done, true variations cannot be ascertained. STEVENS et al. (1972) reported regular seasonal variations in the carbon and oxygen isotopic composition of atmospheric carbon monoxide.

The isotopic composition of atmospheric hydrogen has been determined by BEGEMANN and FRIEDMAN (1959). The δD-values, approximately converted to the SMOW scale, ranged from -200 to $+25‰$. The hydrogen is thus quite similar to meteoric water and may be derived from a photo dissociation of water vapor in the atmosphere. BEGEMANN and FRIEDMAN (1959) postulated that their data refer to mixtures of free atmospheric hydrogen with locally produced industrial hydrogen. Thus it can be concluded that atmospheric hydrogen and the water vapor of the atmosphere are not in thermodynamic equilibrium, as has been pointed out by HARTECK and SUESS (1949).

Sulfur is found as a trace component in the atmosphere where it occurs in aerosoles as sulfate and in the gaseous state as H_2S and SO_2. The sulfur isotopic composition of sulfate in the atmosphere was investigated by OSTLUND (1959), THODE et al. (1961). JENSEN and NAKAI (1961), RAFTER (1965), NAKAI and JENSEN (1967), RAFTER and MIZUTANI (1967), MIZUTANI and RAFTER (1969c), and CORTECCI and LONGINELLI (1970). The latter authors studied the isotopic composition of both sulfur and oxygen in rainwater sulfate.

The following sources of sulfate from rain samples were considered: 1) industrial sulfur, 2) ocean spray sulfate, 3) bacterial sulfur, mostly from tidal flats, and 4) volcanic sulfur.

The results obtained indicate that rainwater sulfate is generally depleted in the heavy isotope with respect to seawater sulfate. MIZUTANI and RAFTER (1969c) concluded that seawater spray was one of the main sources of sulfates in the rain studied, the other main source being industrial activity, which produces SO_2. The contributions of each source varied widely depending mainly upon meteorological conditions. However, about one-half of the samples studied showed a dominant contribution of seawater sulfate.

On the other hand, the results of NAKAI and JENSEN (1967) and CORTECCI and LONGINELLI (1970) suggest that the sulfate in rainwater originates from atmospheric oxidation of SO_2 from industrial activity, without a major contribution from ocean spray.

VII. Biosphere

As used here, the term "biosphere" includes the total sum of living matter — plants, animals, and microorganisms — and the residues of the living matter in the geological environment such as coal and petroleum.

Photosynthesis of carbon dioxide by means of chlorophyll and light is of primary importance for all living matter on earth. The process can be schematically expressed by the following equation:

$$6CO_2 + 6H_2O \rightleftharpoons C_6H_{12}O_6 + 6O_2 .$$

Energy must be supplied if the reaction is to go from left to right. In nature the energy is provided by solar radiation. In this way the radiant energy of sunlight is converted into the chemical energy of organic compounds. The reverse reaction is biological oxidation, or respiration, and the energy thus liberated may appear as heat or work. A fairly close balance exists between photosynthesis and respiration, although over the whole of geological time respiration has been exceeded by photosynthesis, and the energy derived from this was stored in coal and petroleum.

Photosynthesis is of great importance in producing isotope fractionations, not only for carbon but also for hydrogen and oxygen. In general, it is obvious that almost all isotope fractionation due to kinetic effects occurs in the biosphere, as in the cases of the bacterial sulfate reduction, photosynthesis, and respiration.

All organic molecules left over from dead organisms can hardly be preserved without minor or major changes in geological environments. The transformation of biogenic matter to organic matter in sediments involves two stages, a biochemical and a geochemical stage. During the biochemical stage microorganisms are active in reconstituting the organic matter. The biochemical stage is eventually brought to an end when conditions become unsuitable for bacterial activity. Beyond that stage, metamorphism, which results from the action of heat and pressure, is responsible for the further transformation of organic matter.

The origin of coal and petroleum centers around three topics: the nature and composition of the parent organisms, the mode of accumulation of the organic material, and the reactions whereby it was transformed into the end products.

Coal is fossilized plant matter. Depending upon ecological conditions, different plant tissues are converted into various microscopically identifiable components of coal. In addition to these tissues, coal also contains a minor amount of the remains of plant waxes and resins.

Petroleum (frequently also called crude oil) is a naturally occurring complex mixture, composed mainly of hydrocarbons, but also with varying amounts of heterocompounds containing S, N, O, and metallo-organic molecules, such as vanadium- and nickelporphyrins. It appears that most petroleums are similar in their general qualitative compositions and that the differences observed among them are mainly due to differences in relative amounts of certain components.

Although there are, without any doubt, numerous compounds that have been formed more or less directly from biologically produced molecules, the majority of petroleum components are of secondary origin, either decomposition products or products of condensation and polymerization reactions. It should be mentioned that there is strong evidence for the maturation of crude oils, depending on thermal history and age.

1. Carbon

a) Living Matter

The complexities involved in the photosynthetic fixation of carbon have already been briefly discussed. WICKMAN (1952) and CRAIG (1953) were the first to demonstrate that marine plants (except phytoplankton) are about 10‰ heavier in ^{13}C than terrestrial plants. Recent investigations by SMITH and EPSTEIN (1971) have given a more detailed picture. They subdivided the higher plants into two categories:

1) the bulk of the plant kingdom with low $\delta^{13}C$-values (-24 to -34‰),
2) the aquatics — desert and salt marsh plants and tropical grasses with a relatively high ^{13}C-content (-6 to -19‰).

The algae have been put into a separate group with their $\delta^{13}C$-variation range (-12 to -23) lying intermediate between the ranges of the two higher plant groups. The reasons for these differences are not well understood, but as SMITH and EPSTEIN (1971) have shown, plants high in ^{13}C differ from plants low in ^{13}C in anatomy, physiology, biochemistry, and ecology. Adaptations leading to high $^{13}C/^{12}C$ ratios seem to be a response to life under stress.

Animal tissues show the same range as their food supply. PARKER (1964) has analyzed the most abundant biological samples in a shallow marine bay and found that different members of the community had isotopic compositions corresponding to every unit value of ^{13}C between -6 and -17‰, but that the ^{13}C-content was constant for different individuals of the same species within ± 1‰.

DEGENS et al. (1968a) determined the carbon isotope composition of marine phytoplankton cultures that were grown under defined conditions. Their data suggest that algae utilize molecular CO_2 rather than bicarbonate. If molecular CO_2 is highly abundant and in isotopic equilibrium with bicarbonate, the difference between both components may be as high as 28‰. The data of SACKETT et al. (1965), who observed that plankton from arctic regions is depleted in ^{13}C by about 6‰ relative to plankton collected in tropical waters, can be reasonably interpreted as showing that the molecular CO_2 is depleted as temperature and metabolic rate rise.

DEGENS et al. (1968b) have examined the carbon isotopic composition of the major biochemical constituents of marine plankton. Relative to ocean water bicarbonate, the hemicellulose, the proteins, and the pectins are enriched in ^{12}C by about 17‰; cellulose and lignin, by 23‰; and the extractable lipid fraction, by 30‰ (see Fig. 32).

The enrichment in ^{12}C of lipids relative to total plant organic matter has also been shown by SILVERMAN (1964), PARKER (1964), and others.

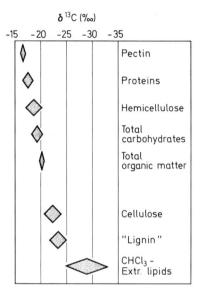

Fig. 32. $\delta^{13}C$ in various biochemical constituents isolated from marine plankton. Diamond-shaped figures represent one unit standard deviation. (After DEGENS *et al.*, 1968 b)

The significant differences among different organic compounds must be considered when the isotopic composition of sedimentary carbon is used as an indicator of the history of sediments.

b) Organic Matter in Sediments

Considering the complex distribution of ^{13}C in living matter (e.g., PARKER, 1964) it is surprising that sedimentary carbon shows a rather narrow variation range. ECKELMANN *et al.* (1962) found in Recent and Tertiary pelagic marine sediments a $^{13}C/^{12}C$ range between -19.3 and $-23.2‰$. Near shore, the contribution by land plants shifts the $\delta^{13}C$ to lighter values (SACKETT and THOMPSON, 1963) (see Fig. 33).

Nonmarine sediments are isotopically similar to land plants (DEEVEY and STUIVER, 1964, and others).

Most ancient Pre-Tertiary marine sediments have $\delta^{13}C$-values similar to present-day terrestrial organic matter, with a mean $\delta^{13}C$ around $-28‰$. This means that ancient marine sedimentary organic matter is generally lighter than Recent sedimentary organic matter. One possible explanation is that the observed enrichment in ^{12}C is a result of diagenetic elimination of the more easily perishable ^{13}C-rich carbohydrate-

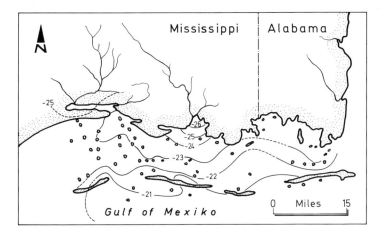

Fig. 33. Isotopic composition of organic carbon in Recent Gulf Coast sediments. (After SACKETT and THOMPSON, 1963)

protein fraction from the marine sediment. Therefore, this feature cannot be used as a criteria for the existence of land-derived organic debris in marine sediments.

c) Petroleum

Numerous $^{13}C/^{12}C$ determinations have been made on petroleum samples, but only relatively few have been published because of economic interest for the oil industry. It should be noted that much of the published data are given relative to the petroleum standard, NBS No. 22 (−29.4‰ relative to PDB).

SILVERMAN and EPSTEIN (1958), SILVERMAN (1964, 1967) demonstrated that the $\delta^{13}C$-values of petroleums are lower than those of the organic matter from which they should derive, but they are similar to the ratios of the lipid fractions, especially of lower plants and animals. The lipids, which are generally 3.5 to 8‰ lighter than the total organic content, are thermally one of the most stable fractions.

On the basis of carbon isotope data, a distinction between predominantly marine and nonmarine oils is frequently possible. The process of petroleum maturation takes place by decomposition, polymerization, and disproportionation reactions of the source molecules. The source molecules during this process form simple molecules relatively rich in hydrogen and ^{12}C on one hand, and more complex molecules containing less hydrogen and ^{12}C on the other hand. This mechanism is schemati-

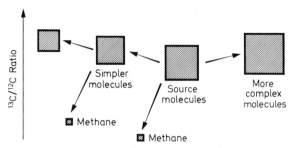

Schematical illustration of petroleum formation

Fig. 34. Relative molecular dimensions are symbolized by the size of the squares, the vertical position of the squares denote relative $^{13}C/^{12}C$ ratios. (From SILVERMAN, 1967)

cally illustrated in Fig. 34. It fits the concept of a protopetroleum composed predominantly of high molecular weight compounds, which could reasonably be composed of lipid components.

A trend toward lighter $\delta^{13}C$ with increasing age (up to Jurassic) has been demonstrated by SILVERMAN (1964). The question of possible variations in the intensity of photosynthesis during geologic history, which may have caused a measurable isotope effect on the isotopic composition of organic matter, is very striking, but difficult to prove (see discussion on p. 89).

d) Natural Gas

Among all natural carbon compounds, methane may be normally most enriched in ^{12}C. Light methane, as low as $-85‰$, has been analyzed by NAKAI (1960), ONANA and DEEVEY (1960), ZARTMAN et al. (1961), WASSERBURG et al. (1963), BOIGK and STAHL (1970), COLOMBO et al. (1966, 1970), and others.

Both isotopic equilibrium reactions and kinetic effects are responsible for the ^{12}C-enrichment in methane. Chemical equilibrium between CO_2 and CH_4 causes an enrichment of $60‰$ in ^{12}C in methane at low temperatures (BOTTINGA, 1969a). Various kinetic effects might be active in the formation of light CH_4. SILVERMAN (1964, 1967) and SACKETT (1968) have shown that the thermal degradation of petroleum compounds and of sedimentary organic material (kerogen) and the accompanying production of methane and other low molecular weight compounds result in an isotope fractionation. ROSENFELD and SILVERMAN (1959) have shown that methane produced during bacterial fermentation of methanole is about $80‰$ lighter than the original host material.

From isotope measurements it is easily demonstrated how light methane can be generated, but it is very complicated to draw conclusions concerning the mode of formation and possible fractionations during migration of methane. Various attempts have been made in this direction; three of these are mentioned. SACKETT (1968) argued that time, temperature, and the structure of parent organic materials are the primary factors regulating the isotopic composition of methane. COLOMBO et al. (1966, 1970) believed that they had demonstrated that during the migration of methane the light isotope is preferentially enriched in the highest zones of a gas field. Contradictory results have been obtained by MAY et al. (1968), who demonstrated a more rapid migration for $^{13}CH_4$ than for $^{12}CH_4$ with $^{12}CH_4$ being better adsorbed than $^{13}CH_4$.

e) Coal

Coals have a $^{13}C/^{12}C$ distribution like that of modern land plants (CRAIG (1953), WICKMAN (1953), COMPSTON (1960), COLOMBO et al. (1970)). Brown coals have been extensively analyzed by FISCHER et al. (1968, 1970). COLOMBO et al. (1970) analyzed coal samples of different coalification grades. No dependence of $\delta^{13}C$ on the facies of the coal has been clearly demonstrated. No change with increasing grade of metamorphism appears to occur.

It is quite remarkable that the process of coal formation, which represents an extensive diagenetic alteration, has not left any measurable fractionation on the $\delta^{13}C$ of coals. A final answer to the problem of a possible isotope fractionation during coalification processes cannot by given yet.

f) "Age Effect" of Carbon Isotope Distribution

This complex problem has been discussed by WELTE (1970). A variation of the carbon isotope composition in the reservoirs

$$CO_{2(atm)} — limestone — biosphere$$

during the geological past might have occurred. The detection of this effect is very complicated because diagenetic alteration might have changed the $\delta^{13}C$ of limestones and organic materials so that they do not represent the primary composition.

From the careful studies of RONOV (1959), TRASK and PATNODE (1939), and WELTE (unpublished data), we know that the distribution of organic carbon in the sediments shows several maxima and minima during earth's history. WELTE (1970) has argued that these irregularities in the distribution are not so much due to a different preservation of organic material, but due to real variations in the intensity of photosynthesis. KEELING (1960) has demonstrated that different photosynthesis rates due to climatic change between summer and winter produce a variation of about 1‰ in the isotopic composition of atmospheric CO_2.

Assuming that the rate of photosynthesis has increased over several million years and is related to a higher accumulation rate of the organic material in the sediments, then a measurable isotope effect in the system

$$CO_{2(atm)} — \text{carbonate rocks} — \text{organic carbon}$$

must have occurred. Such an "age effect" has not been proven with certainty, but some indications favor it.

Several authors (e.g., COMPSTON, 1960; KEITH and WEBER, 1964; WEBER, 1967) have tried to find an "age effect" in limestones. However, because of possible diagenetic alterations, limestones are not very suitable. It also seems almost impossible to demonstrate an age effect on crude oils, because of the mobility of oils and their variability in composition.

Kerogen-like material present in shales seems more suitable for an analysis of this type (WELTE, 1970).

Most interesting in this connection is the question of the carbon isotopic composition during early Precambrian time. From the available data we must conclude that at least during late Precambrian time the $^{13}C/^{12}C$ ratios were not too different from today.

2. Hydrogen

Plants remove hydrogen from water and transfer it to organic compounds. Although many organic hydrogen atoms are exchangeable with environmental water, once a carbon-hydrogen bond is formed in an organism, the hydrogen seems to be no longer readily exchanged. Studies on the natural abundance of deuterium and hydrogen in organic matter done by BOKHOVEN and THEEUWEN (1956) and RITTENBERG (1963) made no attempt to relate the D/H ratios from organic matter to the hydrogen source, e.g., environmental water.

More recent determinations by ZBOROWSKI et al. (1967), SCHIEGL and VOGEL (1970), and SMITH and EPSTEIN (1970) showed that the plants are depleted in deuterium relative to the environmental water.

SMITH and EPSTEIN (1970) gave fractionation factors from water hydrogen to whole plant hydrogen from 1.008 to 1.085, with the average value of 1.055. Fractionation factors from whole plant to lipid ranged from 1.060 to 1.141, with an average value of 1.092.

SCHIEGL and VOGEL (1970) attempted to correlate organic D/H ratios with waters precipitated from the atmosphere. They observed three different fractionation effects:

1) a depletion in the deuterium content of several percent during the conversion of water to organic matter in living plants,

2) a decrease in the D-concentration going from living plants to coal, and

3) a further depletion in deuterium in the methane being generated from coal and oil.

Note the similarity in the fractionation trend as discussed for the carbon isotope composition.

3. Sulfur

At present only a limited number of measurements on $^{34}S/^{32}S$ ratios of biological material are available.

MEKHIYEVA and PANKINA (1968) have demonstrated that sulfur of aquatic plants from a given water is slightly lighter than the sulfur of the dissolved sulfate. The same results have been obtained by KAPLAN et al. (1963) for marine organisms — plants and animals.

Several studies have been made on the $^{34}S/^{32}S$ ratios in natural crude oil and associated materials, e.g., THODE et al. (1958), HARRISON and THODE (1958), PANKINA and MEKHIYEVA (1964), THODE and MONSTER (1965), and THODE and MONSTER (1970), THODE and REES (1970).

Sulfur isotope studies of petroleum and related sulfates and sulfides can give information concerning the possible origin of petroleum sulfur and concerning the environment in which the petroleum was formed. Most oils are formed in a marine environment, and it seems reasonable to assume that ocean water sulfate is the main source of sulfur in petroleum. Since seawater sulfate is rapidly reduced by bacteria in the shallow muds in contact with the sea, it is very likely that it would be this reduced sulfur that would finally be incorporated in the petroleum (sulfur originating from the organic primary material is only of minor importance). In general, the $\delta^{34}S$-value for petroleum is depleted by $\sim 15‰$ relative to associated evaporite sulfates, which would be in agreement with a bacterial reduction of ocean water sulfate.

In general, single large oil pools are uniform in isotopic composition. In contrast, oil pools in reservoir rocks of different geologic ages show a wide range of isotope ratios ($\delta^{34}S$ varies by about 45‰).

The pattern of sulfur isotopes in marine oils is very different from that in nonmarine oils. For marine oils the $\delta^{34}S$-values are related to those of the contemporaneous oceans. Nonmarine oils have no relation to the isotope ratios for contemperaneous oceans.

These results, together with the fact that hydrogen sulfide gas derived from many of these oils has similar isotope ratios, (THODE et al., 1958) indicate that there is little fractionation of the sulfur isotopes in the maturation or degassing of oil during which sulfur is lost. Thus, the sulfur isotope ratio for oil seems to give information about the sulfur isotope content of the source sulfur at the time of petroleum formation.

VIII. Sedimentary Rocks

Sediments are the weathering products and residues of magmatic, metamorphic and sedimentary rocks after transport and accumulation in water and air. The recycling of sediments, incorporated into new sedimentary rocks, has been increased during the earth's history.

Classification of sedimentary rocks is based on easily recognizable characteristics that reflect something about the mode of transport and the environment of deposition. It is customary to consider sedimentary rocks in two categories: *clastic* and *chemical.* Transported fragmental debris of all kinds — sands, gravel, shell fragments — make up the clastic component of the rock. Inorganic precipitates from water (e.g., gypsum and silica) obviously belong to the chemical category. But the chemical components also include biogenic material extracted from waters or partly from air and secreted as skeletons of living organisms. One of the most serious problems in this classification is the possible mingling of the two classes.

1. Silicates

Chemical weathering of high-temperature silicates proceeds mainly by reactions with water, carbon dioxide, and oxygen. Compared with the composition of the primary magmatic rocks, average sediments are distinctly higher in water, carbon, and chlorine, and also in the ratio of ferric to ferrous iron.

The $^{18}O/^{16}O$ and D/H ratios of weathering products will be mainly controlled by 1) the isotopic composition of the coexisting waters, 2) by the isotopic fractionation factors between the water involved and each of the minerals formed during weathering and 3) by the temperature (SAVIN and EPSTEIN, 1970a; LAWRENCE and TAYLOR, 1971). Because the weathering normally involves large amounts of meteoric water relative to the amount of parent rock, the isotopic composition of the parent rock has little influence on the isotopic composition of the weathering products. LAWRENCE and TAYLOR (1971) were able to confirm that the isotopic composition of weathered rocks mainly reflects the isotopic variations of meteoric waters. From this follows that clay minerals which originated in fresh waters or underwent diagenetic isotopic exchange have isotope ratios that are smaller than those minerals whose isotopic compositions have been established in a marine environment.

Most of the soils found on earth have $\delta^{18}O$-values from $+15$ to $+25‰$ and δD-values from $-30‰$ to $-100‰$. Soils rich in gibbsite or other Al-Fe hydoxides are isotopically lighter in ^{18}O than clay-rich soils. SAVIN and EPSTEIN (1970a) and LAWRENCE and TAYLOR (1971) estimated that kaolinites and montmorillonites are about 27‰ richer in ^{18}O

than the coexisting water, whereas gibbsites are only 18‰ richer in ^{18}O. On the other hand, all the clay minerals and hydroxides are depleted in deuterium relative to the coexisting water: kaolinite and montmorillonite: ∼ 30‰, gibbsite: ∼ 15‰.

LAWRENCE and TAYLOR (1970) have used the δD- and δ^{18}O-values of ancient kaolinites to determine δD and δ^{18}O of ancient meteoric waters. Their results suggest that the isotopic composition of Mid-Tertiary waters is similar to the present-day composition.

The proportion of detrital to authigenic minerals of argillaceous rocks has been extensively debated. For some time authigenic clay mineral formation has been assumed to produce the largest fraction of modern sediments. But recent experimental results have demonstrated that the predominant fraction of clays and shales must be of detrital origin.

^{18}O/^{16}O ratios of modern ocean sediment core samples indicate that generally the authigenic components may be distinguished from detrital components (authigenic minerals are generally 10 to 30‰ heavier in ^{18}O than detrital minerals of high-temperature origin, SAVIN and EPSTEIN, 1970b).

^{18}O/^{16}O ratios of detrital minerals appear to reflect the provenance and mode of origin. Detrital quartz, for instance, seems to be resistant to weathering, and it will retain its original ^{18}O-content as established in the parent rocks. REX et al. (1969) for instance found that the ^{18}O/^{16}O ratio of quartz isolated from Hawaiian soils, Pacific sediments, and tropospheric dusts are remarkably uniform. The authors suggested a common aeolian origin of this quartz from continental land masses.

On the other hand the δ^{18}O-values of authigenic minerals suggest formation in the marine environment under or near to isotopic equilibrium conditions. The existence of authigenic feldspars of sedimentary or diagenetic origin has long been recognized. SAVIN and EPSTEIN (1970c) analyzed feldspar grains consisting of secondary overgrowths on detrital cores and found a striking difference of approximately 10‰ between the isotopic composition of the core and that of the rim. The typical igneous δ-value obtained for the core indicates that it did not exchange during the formation of the authigenic feldspar.

SAVIN and EPSTEIN (1970b) extended their investigations on pure clay minerals to the more complicated fine grained ocean sediments. The isotope data on whole-rock ocean sediments cannot be interpreted without a knowledge of their chemical and mineralogical composition. Some of the major minerals present in the carbonate-free fractions of ocean sediments are the clay minerals whose isotopic compositions have been discussed before, quartz, feldspar and the iron and manganese oxides. To calculate the average oxygen isotope composition of sediments, we have to know the quantitative mineralogical and isotope composition of the pure phases. Quartz in ocean sediments seems to be mostly of detrital origin. SAVIN and EPSTEIN (1970b) assigned a $\delta^{18}O$-value of $+18‰$. Feldspar, too, appears to be mostly of detrital origin and probably has a δ-value between $+7$ and $+10‰$. Very few results have been obtained on the $^{18}O/^{16}O$ ratios of iron and manganese oxides. SAVIN and EPSTEIN (1970b) assumed that the isotopic composition of the oxides are similar to the manganese nodule analyzed ($\delta^{18}O$: $+15‰$). With these data, the average composition of marine shales being relatively similar in mineralogy and chemistry to recent ocean sediments can be calculated. SAVIN and EPSTEIN (1970b) found a variation range of $+14$ to $+19‰$ for the oxygen isotope composition in shales.

δD-values on the carbonate-free fractions of ocean core samples from nearly all over the world range from -55 to $-87‰$ (SAVIN and EPSTEIN, 1970b). There is only inconclusive evidence for hydrogen isotopic exchange between these samples and ocean water.

2. Carbonates

a) Marine Carbonates

The question of the isotopic composition of carbonates is intimately linked with the term "paleotemperature".

In 1946 UREY presented a paper concerning the thermodynamics of isotopic systems and suggested that variations in precipitation temperatures of calcium carbonate from water should lead to measurable variations in the $^{18}O/^{16}O$ ratio of the calcium carbonate. He postulated that the determination of temperatures of the ancient oceans should be possible, in principle, by measuring the ^{18}O-content of fossil calcite shells. (One can reasonably assume that only the $^{18}O/^{16}O$ ratio of the carbonate will be temperature dependent, because the amount of water in the oceans is so much greater than the amount of dissolved carbonate.)

At known water temperatures, EPSTEIN et al. (1953) have shown experimentally that marine organisms such as abalones do secret calcareous shells in chemical equilibrium. EPSTEIN and coworkers obtained the following empirical relationship, slightly modified by CRAIG (1965)

$$t° C = 16.9 — 4.2\varDelta + 0.13\varDelta^2$$

where $\varDelta$ is the per mil difference between CO_2 derived from carbonate by reaction with H_3PO_4 at $25°$ C and CO_2 equilibrated with the water at $25°$ C from which the carbonate was deposited.

Three problems make paleotemperature determinations rather uncertain:

1) the unknown ^{18}O-content of ancient oceans,
2) metabolic effects on carbonate precipitation,

3) the isotopic preservation of primary oxygen in the carbonates.

1) As has already been mentioned, it may be concluded from several arguments that ocean water has had a constant isotopic composition within small limits during geological history. Glacial periods during the history of the earth are, however, excluded.

We may, therefore, reasonably assume that ancient ocean water had a $^{18}O/^{16}O$ ratio similar to that of today. However, a crucial point is the question of "paleosalinities". We must know if the organism to be analyzed has lived in ocean water of 35‰ salinity. Ocean water of higher salinities has a higher ^{18}O-content, because during evaporation ^{16}O is preferentially concentrated in the vapor phase; ocean water of lower salinities has a lower ^{18}O-content, because it is diluted by fresh waters. EPSTEIN and MAYEDA (1953) estimate that variations in salinities of one unit in 35‰ salinity would be accompanied by 1° C error in temperature determinations in a nonglacial period of the history of the earth.

2) Some organisms, e.g., mollusks (LLOYD, 1964) and brachiopods (LOWENSTAM, 1961), apparently deposit calcite or aragonite in isotope equilibrium with ocean water. In contrast, other organisms, e.g., echinoderms, asteroidea, ophiuroidea, and crinoidea (WEBER and RAUP, 1966a, b; WEBER, 1968) do not precipitate their carbonates in equilibrium with their environment. These observations are accounted for by an isotope-exchange reaction between respiratory CO_2 and dissolved bicarbonate at or near the site of skeletal deposition. The fact that metabolic fractionation occurs in many organisms is most important in respect to the problem of paleotemperature determinations with the help of ammonites and belemnites and other species (TOURTELOT and RYE, 1969; SPAETH et al., 1971).

A knowledge of the ecologic behavior of shell-secreting organisms is essential. If still-existing species are used for thermometry, the assumption must be made that their depth habitats have not changed with time. EICHLER and RISTEDT (1966) reported variations in the $^{18}O/^{16}O$ ratio of the early shell and septa of two Nautilus specimens, which are in part ascribed to migrations from warmer to cooler water after a certain stage of development. How complicated these processes are can be demonstrated in the measured difference between right- and left-coiled foraminifera, observed by LONGINELLI and TONGIORGI (1964). One explanation might be the preferential growth of one form or the other within a certain temperature range.

Another important point is the question whether the calcium-carbonate-secreting organisms grow shells only during a portion of the local temperature range or throughout the entire range. EPSTEIN and LOWENSTAM (1953) have shown that growth of skeletons of most species does not take place during the entire year. The majority of the pelecypods, for instance, seems to grow primarily in the warm temperature range, whereas the gastropods show winter as well summer growth. Some forms retain shell growths at low temperatures.

3) The third point seems to be the most serious one. The isotopic composition of oxygen in an aragonite or calcite shell will remain unchanged until the shell material dissolves and recrystallizes during diagenesis. Some of the criteria by which unaltered samples might be recognized have been discussed by LOWENSTAM (1961). But the problem of how to prove the preservation is still unsolved.

Paleotemperature techniques have been moderately successful when applied in the climatic history of the Pleistocene. The Pleistocene epoch appears to have been characterized by alternation between climates similar to those experienced today and colder climates.

EMILIANI (1955, 1958, 1966) has measured isotopic variations of foraminifera of ocean sediments and observed eight cyclic isotopic fluctuations in the foraminifera of cores representing about 425000 years. As has already been mentioned, one of the chief unknown quantities in the determination of past temperatures is the isotopic composition of the ocean water in the geological past. SHACKLETON (1967, 1968, 1969) estimated a glacial-interglacial change in the isotopic composition of ocean water of approximately 1.0 to 1.4‰. SHACKLETON (1968) assumed that about one-third of the total isotopic variation observed in Caribbean cores by EMILIANI should be attributed to temperature change and two-thirds to isotopic change in the water.

b) Limestones

Isotopic data on several hundred limestone samples have been reported so far in the literature, the majority of which have a $\delta^{13}C$-content close to zero. KEITH and WEBER (1964) have stated that the average carbon isotopic composition of marine limestones appears to be moderately constant from Cambrian through Tertiary time.

During diagenesis, marine carbonates generally become lighter in carbon isotope composition (BAERTSCHI, 1957; GROSS, 1964). Submarine alteration, however, may generate heavier cement than primary bioclastic carbonate (FRIEDMAN, 1964).

During early diagenesis and the following geological history most limestones also re-equilibrate their ^{18}O-content with the surrounding pore fluids, which are generally different from ocean water. Most carbonates become lighter with increasing age (DEGENS and EPSTEIN, 1964; KEITH and WEBER, 1964). Variations with age in the isotopic composition of marine and freshwater limestones are shown in Fig. 35.

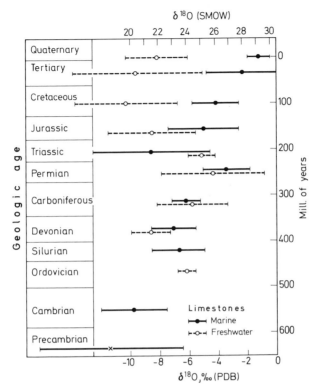

Fig. 35. Variations of $\delta^{18}O$-values of marine and freshwater limestones with time. (After KEITH and WEBER, 1964)

c) Freshwater Carbonates

Freshwater carbonates are generally lighter and show a much broader variation range in carbon and oxygen isotope composition than marine carbonates, because fresh water is generally lighter and more variable in isotope composition than ocean water. For instance, freshwater limestones are generally lighter in carbon isotope composition than marine limestones, which may be attributed to a variable contribution of organically derived CO_2 generated mainly in soils (VOGEL, 1959; CLAYTON and DEGENS, 1959; KEITH and ANDERSON, 1963; KEITH et al., 1963; KEITH and WEBER, 1964) (see also the footnote on p. 24).

The formation of some carbonates is related to CO_2 produced during microbiological oxidation of the organic matter embedded in the sediments. These carbonates may show anomalously low $^{13}C/^{12}C$ ratios. Under this category are carbonates associated with elemental sulfur (CHE-

NEY and JENSEN, 1965) and carbonate concretions quite often found in clays and shales (HODGSON, 1966; MURATA et al., 1967; HOEFS, 1970).

Carbon and oxygen isotope composition of marine and freshwater limestones have been used as environmental indicators (CLAYTON and DEGENS, 1959; KEITH and WEBER, 1964).

The environmental interpretation of sedimentary rocks is one of the principle objectives of sedimentary petrology. In the past such interpretations have been based primarily on certain specific fossils and on certain trace element contents — criteria that have obvious limitations. Therefore, it is quite natural to study the application of stable isotope measurements as possible environmental indicators (i.e., CLAYTON and DEGENS, 1959; KEITH and WEBER, 1964).

Since fresh water is in general depleted in ^{18}O relative to ocean water and more variable in the $^{13}C/^{12}C$ ratio due to a possible relatively high contribution of organic derived CO_2, it should be possible, in principle, to differentiate between the marine, brackish, and freshwater environments.

KEITH and WEBER (1964) gave the following equation for Jurassic and younger samples to discriminate between marine and freshwater limestones

$$Z = a(\delta^{13}C + 50) + b(\delta^{18}O + 50)$$

in which a and b are 2.048 and 0.498, respectively. Limestones with a Z-value above 120 would be classified as marine; those with a Z-value below 120, as freshwater types.

However, it must be taken into account that the isotopic composition of carbonate in a rock today depends not only on the environment at the time of deposition, but also in isotopic exchange processes since deposition. Another limitation is the necessity to demonstrate that the carbonate in sediments is of sedimentary origin and not contaminated by either detrital or post-sedimentary processes. It should be further emphasized that fresh water is so variable in isotopic composition that some of it may be similar to ocean water.

To summarize, it seems obvious that isotopic criteria should be regarded as additional parameters, supplemental to paleontologic and petrographic criteria, for determining the depositional environment of sedimentary rocks.

Contradictory conclusions have been reached about the relationship between the $\delta^{13}C$ and $\delta^{18}O$-values of speleothems and climatic variations. Speleothems (MOORE, 1952), such as stalactites and stalagmites, are formed when calcium carbonate is precipitated from solutions seeping into limestone caves. The possibility of obtaining paleoclimatic data from speleothems has recently stimulated interest in the distribution of carbon and oxygen isotopes in the speleothem carbonates (GALIMOV and GRINENKO, 1965; FORNACA-FINALDI et al., 1968; HENDY and WILSON, 1968; DUPLESSY et al., 1969; and HENDY, 1971).

The isotopic composition of the calcite deposited on speleothems will depend both on the manner in which the limestone carbonate was dissolved by the ground waters and the manner in which the precipitation took place. Three different modes of deposition may be distinguished (after HENDY, 1971):

"1) If isotopic equilibrium is maintained between HCO_3^- and CO_2, the calcite precipitated will be in isotopic equilibrium with the water, and variations in $^{18}O/^{16}O$ will depend on climate alone. This will occur if the loss of CO_2 from solution is slow.

2) If loss of CO_2 is rapid, a kinetic fractionation will occur between HCO_3^- and $CO_{2(aq)}$, and the calcite precipitated will show a simultaneous enrichment in ^{13}C and ^{18}O.

3) If evaporation of water occurs, the calcite precipitated will be enriched in ^{18}O. Speleothems deposited under conditions of kinetic loss of CO_2 or evaporation of water cannote be used to give paleoclimatic data."

d) Dolomites

The formation of dolomites is still a widely debated question in the geochemistry of carbonate sediments. This applies very much to the results of stable isotope determinations, where despite relatively large numbers of carbon and oxygen isotope determinations, no firm conclusions appear possible (BAUSCH and HOEFS, 1972; CLAYTON et al., 1968a and b; DEGENS and EPSTEIN, 1964; DEUSER, 1970; EPSTEIN et al., 1964; FONTES et al., 1970; FRITZ and SMITH, 1970; FRITZ, 1971; GROSS and TRACEY, 1966; HALL and FRIEDMAN, 1969; MURATA et al., 1967; NORTHROP and CLAYTON, 1966; O'NEIL and EPSTEIN, 1966b; SHEPPARD and SCHWARCZ, 1970; TAN and HUDSON, 1971; WEBER, 1964, 1965).

At low temperatures the isotopic fractionation of oxygen and carbon in the dolomite-water system is not very well known. Extrapolations from high-temperature experiments give a 5.0 to 7.0‰ enrichment in ^{18}O in dolomite compared to syngenetic calcite, whereas the carbon isotope fractionation is much smaller, with about 2.5‰ enrichment of ^{13}C in dolomite. The problem, however, is that only very few examples with this theoretically expected equilibrium factor have been found in nature. Most dolomites analyzed so far are either slightly heavier or equal, although some are even lighter in isotopic composition than coexisting calcite (BAUSCH and HOEFS, 1972).

One way of explaining this discrepancy is to argue that the oxygen equilibrium fractionation factor is smaller than 5 to 7‰ (FRIEDMAN and HALL, 1963; HALL and FRIEDMAN, 1969). Another possibility is that if the dolomite was deposited as metastable protodolomite, the protodo-lomite-water fractionation should be considerably smaller than the do-lomite-water fractionation (FRITZ and SMITH, 1970; TARUTANI et al., 1969).

However, the question of isotopic equilibrium might not be relevant in this connection. If we assume that dolomite is always formed by the replacement of calcium carbonate, then the oxygen isotope composition of the dolomitizing solutions should be of great importance. Dolomites formed at an early diagenetic stage should be similar or slightly enriched in ^{18}O relative to the precursor calcite, since during early diagenesis pore solutions should be similar to ocean water. Late diagenetic solutions should be more variable in isotopic composition; therefore, the late diagenetic dolomites should be more variable, too (BAUSCH and HOEFS, 1972).

Appreciable variations in $^{13}C/^{12}C$ ratios have also been found in do-lomites (from $+21$ to $-64‰$, MURATA et al., 1967; DEUSER, 1970). The light values are directly related to the oxidation of organic matter; the heavy ones could be indirectly related (through equilibration of light methane with heavy carbon dioxide). SPOTTS and SILVERMAN (1966) were able to demonstrate the production of small dolomite crystals around oil droplets, which were formed during the oxidation of organic matter.

3. Phosphates

As has already been pointed out by UREY et al. (1951), the develop-ment of another temperature scale using calcium carbonate and another oxygen compound precipitated by organisms, i.e., phosphates, would permit a temperature scale independent of the oxygen isotopic composi-tion of ocean waters.

LONGINELLI (1965, 1966) analyzed co-precipitated carbonate and phosphate and used the carbonate calibration to obtain a tentative cali-bration of the phosphate-water paleothermometer. LONGINELLI and NUTI (1968a) suggested that fossil marine organisms containing phos-phatic material preserved the original isotopic composition moderately well and that the phosphatic material was deposited under isotopic equi-librium conditions. However, in this connection it must be considered that phosphorous is a key element in all biological processes. The phos-phate metabolic cycle is so complicated that it seems impossible to exclude an isotope exchange due to "life processes".

LONGINELLI and NUTI (1968b) measured the $^{18}O/^{16}O$ ratios of phos-phorites from marine formations of different geological ages. They inter-preted their results as showing that marine phosphorites of Upper Ter-tiary age were probably formed under isotopic equilibrium conditions.

4. Sedimentary Sulfides

Due to the activity of sulfate-reducing bacteria (producing light sulfide and heavy residual sulfate) most of the sulfur isotope fractionation takes place in the uppermost layers of the muds of shallow sea basins and of tidal flats. THODE et al. (1954), FEELY and KULP (1957), THODE et al. (1960), VINOGRADOV et al. (1962), KAPLAN et al. (1963), NAKAI and JENSEN (1964), HARTMANN and NIELSEN (1969), and others found that in various natural environments enrichments occurred in ^{32}S from 15 to 62‰ for sulfides relative to associated sulfates.

Sedimentary sulfides should normally have a δ^{34}S-value somewhere between -10 and -30‰, although numerous examples exist where the sedimentary sulfides show rather heavy δ-values.

Sulfur isotope ratios in sediments are sensitive indicators of certain variable conditions in the microenvironment of a given sedimentation region. VINOGRADOV et al. (1962) found in Recent Black Sea sediments δ^{34}S-values between -20 and -36‰ for the various sulfur compounds (pyrite, H$_2$S, elemental sulfur) in the H$_2$S-containing region. More positive values between -6 and -21‰ have been found in sediments from the oxygen-containing zone. KAPLAN et al. (1963) report δ^{34}S-values between -16.6 and 3.1‰. The organic sulfur from decaying organic matter apparently plays only a small role in the sulfur economy. HARTMANN and NIELSEN (1969) demonstrated that as sediment depth increases, the pore water sulfate shows increasing δ^{34}S-values and that the ^{34}S/^{32}S ratios of the pore water sulfide increases parallel to the sulfate sulfur, with a nearly constant δ-difference. These examples should make it clear that even under similar sedimentation conditions, the ^{34}S-contents of sediments can vary considerably.

The isotopic composition of sedimentary sulfates (evaporites) is discussed under the isotopic composition of ocean water.

5. Native Sulfur Deposits

The large native sulfur deposits found on earth, e.g., Sicily, Louisiana, and Texas, are always associated with gypsum and anhydrite and are usually encountered within the cap rock of salt domes. THODE et al. (1954), FEELY and KULP (1957), DESSAU et al. (1962), SCHNEIDER and NIELSEN (1965), and DAVIS and KIRKLAND (1970) analyzed sulfur isotope ratios of native sulfur of this type. The isotope measurements have provided overwhelming evidence that the native sulfur is formed by the reduction of sulfate by sulfate-reducing bacteria (the normally associated petroleum provides a source of carbon and the energy for metabolism).

Together with the sulfur there is often associated a calcitic limestone matrix, which was found to be depleted in ^{13}C — which is again convincing evidence for an organic origin of the carbonate and a bacterial origin of the sulfur.

IX. Metamorphic Rocks

Metamorphic rocks are mineralogically transformed sedimentary or igneous rocks. The temperature range of metamorphism is generally between that of the sedimentary environment and that at which magma is generated by rock fusion, or even above this temperature (granulite facies). There is sometimes an overlap with low-temperature diagenetic processes, and at the other extreme, metamorphism overlaps conditions under which the more fusible minerals begin to melt in the presence of water (migmatite).

Metamorphism occurs in essentially solid systems, with little or no contribution from silicate-rich melts. However, a water vapor and carbon dioxide gas phase plays an essential role.

As has already been demonstrated, every major class of rock has a characteristic range of isotope composition. For the oxygen isotope composition, VERHOOGEN et al. (1970) have put it the following way: "In igneous rocks it is determined by the source of the primary magma and by isotopic fractionation during fractional crystallization. The oxygen of sedimentary rocks has a composition that reflects a state of equilibrium between weathering products and meteoric or ocean water ... The isotopic composition of oxygen in metamorphic rocks gives a good indication of the extent to which externally supplied water may have affected the original composition inherited from the parent sedimentary or igneous rock. (For example, amphibolites formed by regional metamorphism of basalt have significantly higher ^{18}O-contents than most basalts. Pelitic schists tend to give values somewhat lower than shales. What seems to be implied is isotopic exchange with an external reservoir of water.)"

Regarding the isotopic composition of metamorphic rocks, two problems are of special interest: First the question of whether isotopic equilibrium has been attained during metamorphism; second, the determination of a "temperature of formation".

Fractionations among coexisting minerals commonly indicate that isotopic equilibrium has often been nearly attained in the mineral assemblages during metamorphism. A good example of large-scale oxygen isotope equilibrium over several thousand km^2 was observed by TAYLOR (1964) in the Adirondack Mountains, New York. However, sometimes there is obviously a lack of isotope equilibrium. PERRY and BONNI-

CHSEN (1966) and ANDERSON (1967) have described heterogenous metamorphic rocks in which detectable isotopic disequilibrium occurs over a
distance of a few centimeters.

The possibility of using oxygen isotope fractionations as a geothermometer in metamorphic rocks was demonstrated by CLAYTON and
EPSTEIN (1958), and has since been applied by JAMES and CLAYTON
(1962), SCHWARCZ (1966), GARLICK and EPSTEIN (1967), SHEPPARD and
SCHWARCZ (1970), DEVEREUX (1968), TAYLOR and COLEMAN (1968),
CLAYTON et al. (1968 c), SHIEH and TAYLOR (1969 a, b), SCHWARCZ et al.
(1970), WILSON et al., (1970), and others.

Since GARLICK and EPSTEIN (1967) found that temperature as inferred from the quartz-magnetite geothermometer agreed in general with
the usual geologic estimates based on experimental studies of silicate
assemblages, the quartz-magnetite pair is generally regarded as the most
suitable for the determination of metamorphic isotopic temperatures.
However, as SCHWARCZ et al. (1970) have shown although the Δ-differ-
ence of a mineral pair general decreases with metamorphic grade there is
a significant spread within each zone, which may be governed by the
following factors:

1) The variations in the temperatures are due to compensating differences of other variables besides temperature (e.g., chemical composition,
oxygen fugacity, p_{H_2O}).

2) The ideal isotopic geothermometer should record only the maximum temperature of metamorphism. However, a mineral pair may con-
tinue to re-equilibrate to some temperature below the maximum temperature of metamorphism.

3) Some minerals may have formed or re-equilibrated during discrete metamorphic events later than, and of a lower temperature than,
the maximum metamorphic temperature.

Assuming that the spread in Δ-differences of given mineral pairs are
due to retrograde exchange, we can take the minimum fractionations in
each zone to represent most closely the maximum temperatures of metamorphism.

Quartz-mica fractionations do not appear to be a promising geothermometer. CLAYTON (manuscript in preparation) has shown in a large
number of igneous and high-temperature metamorphic rocks, that assemblages involving micas give rarely concordant temperatures and that
quartz-muscovite fractionations usually fall in the range 2.4 to 3.4‰,
irrespective of rock type.

SCHWARCZ et al. (1970) have attempted to test a quartz and calcite
geothermometer. They find that "fractionation values decrease with increasing metamorphic grade up to the garnet zone but level off or increase at higher grades, showing that at least at these higher grades the

isotopic fractionations do not preserve a record of the highest temperature of metamorphism ($T_{(max)}$). Therefore, there has been retrogression down to some temperature below $T_{(max)}$ at which equilibrium may again be frozen in. This is probably because subsequent to reaching $T_{(max)}$, some minerals in the rock continued to exchange with one another, down to some lower temperature."

Another important point of oxygen isotope determinations on metamorphic rocks is that there is evidence for interaction between the metamorphic mineral assemblages and a fluid phase permeating the rocks at the time of metamorphism. TAYLOR and COLEMAN (1968) have pointed out that the result of this interaction is that any mineral in equilibrium with an isotopically homogeneous pore fluid would attain the same $^{18}O/^{16}O$ ratio irrespective of rock type as long as the temperature is constant. Quartz for instance should attain the same oxygen isotope composition in all adjoining meta-sedimentary or meta-igneous rocks, because of equilibration with the same pore fluid.

The effect of dehydration reactions upon the oxygen isotope composition of pelitic rocks can be studied by examining the variations of the ^{18}O-content across metamorphic zones. SHIEH and TAYLOR (1969a) have found that dehydration reactions would not produce any appreciable oxygen isotopic changes at least during contact metamorphism.

Metamorphism has been classified into several different types, two most important of which are:

1) *Regional* metamorphism, occurring on a regional scale in areas up to thousands of square kilometers.

2) *Contact* metamorphism, taking place in heated rocks bordering magmatic intrusions.

1. Regional Metamorphism

When considering regional metamorphic rocks, the most striking feature is that the $^{18}O/^{16}O$ ratios of pelitic rocks tend to decrease with increasing metamorphic grade. The $\delta^{18}O$-value in pelitic rocks of low grade (chlorite zone) is close to the values typical of shales (15 to 18‰). It decreases more or less steadily through the biotite, garnet, and staurolite-kyanite zones to values in the range 9 to 13‰ for high-grade sillimanite schists. A comparison of $\delta^{18}O$-values of regional with contact metamorphic rocks is shown in Fig. 36.

2. Contact Metamorphism

Because the oxygen isotopic composition of igneous rocks is quite different from that of sedimentary and low-grade metamorphic rocks, studies on the variations of oxygen isotopes in the vicinity of an intrusive

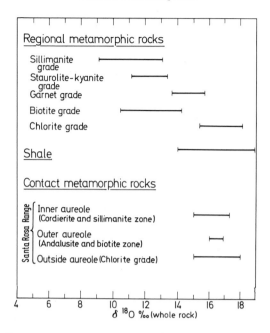

Fig. 36. O^{18}/O^{10} ratios of pelitic rocks plotted against metamorphic grade for both regional and contact metamorphism. (After SHIEH and TAYLOR, 1969a)

contact offer the possibility to investigate the extent of isotopic exchange between the intrusive and its country rock.

In contrast to regional metamorphic rocks, the $^{18}O/^{16}O$ ratios in contact metamorphic rocks remain exceedingly constant throughout the entire contact aureole (see Fig. 36). Somewhat surprisingly, $\delta^{18}O$-values in pelitic hornfels of a number of contact aureoles have been found to remain very much the same as in the parent sedimentary material, except within a meter or so of plutonic contacts (SHIEH and TAYLOR, 1969a).

The fact that only a very narrow zone of country rock undergoes isotopic exchange with the intrusive implies that horizontal outward movement of water into the contact metamorphic aureole is almost negligible during crystallization of the pluton.

This indicates that most of the water involved in contact metamorphism is derived from the country rocks themselves rather than from the intrusive.

Based on oxygen isotope fractionation calibration curves, SHIEH and TAYLOR (1969a) estimated the contact temperatures of granitic plutons and country rock to be between 525° C and 625° C, which agrees reasonably well with temperatures estimated from heat flow calculations, from mineral parageneses, and from melting experiments.

Since metamorphic reactions commonly involve hydration and dehydration, water must be able to diffuse into or out of the reacting system. If, for example, the metasedimentary rocks have undergone isotopic exchange with significant quantities of pore-fluids in communication with or derived from plutonic igneous rocks during metamorphism, one should expect a decrease in $^{18}O/^{16}O$ ratios with increasing metamorphic grade. If metamorphism involves only simple loss of water through dehydration reactions, then the $^{18}O/^{16}O$ ratios of metasedimentary rocks should increase during progressive metamorphism.

However, as we have demonstrated, contact metamorphic rocks seem to preserve the original δ-values of the parent sedimentary materials. This should mean that they themselves have supplied the metamorphic water.

In the case of regional metamorphism the situation seems to be entirely different. Only exchange with an extensive external oxygen reservoir can account for systematic, observed ^{18}O-changes. SHIEH and TAYLOR (1969a) postulate exchange with fluids that are ultimately derived from deep-seated plutonic igneous bodies. However, this view leaves many questions unanswered.

Summarizing, regional metamorphism usually involves exchange with an external oxygen reservoir, whereas contact metamorphism generally does not.

In addition to the dominating importance of determining "temperatures of formation" of metamorphic rocks, there are broad possibilities for applying $^{18}O/^{16}O$ studies in metamorphic petrology. Some of these possibilities are cited in the following.

Since metamorphic rocks are transformed sedimentary or igneous rocks, both of which show very different initial isotopic compositions, $^{18}O/^{16}O$ studies might be useful in distinguishing ortho-rocks from para-rocks based on the idea that the effects of initial isotopic differences in the precursor rocks have been to some extent retained (SCHWARCZ and CLAYTON, 1965). VOGEL and GARLICK (1970) determined $\delta^{18}O$-values in metamorphic eclogites and found two types: those with anomalously low δ-values and those with high δ-values. They concluded that the isotopically light eclogites originated from basaltic rocks that interacted with light meteoric waters at high temperatures, and that the isotopically heavy eclogites are possibly derived from dolomitic pelites.

TAYLOR and COLEMAN (1968) have shown that glaucophane-bearing metamorphic rocks form over a very wide temperature range probably from 200 to 550° C. These authors suggested that blue-schist-facies rocks may be subdivided into two separate facies: a low-temperature lawsonite-aragonite facies and a higher-temperature, epidote-rutile blueschist facies.

HOEFS and EPSTEIN (1969) analyzed some migmatites, which seemed to have "mixed values" between igneous and metamorphic rocks. These values were interpreted as an indication of an out-of-equilibrium state.

WENNER and TAYLOR (1971) calculated temperatures of serpentinization of ultramafic rocks. They argued that many lizardite-chrysotile serpentines must have formed at very low temperatures compared to the antigorites. They suggested that the lizardite-chrysotile serpentinization in most alpine ultramafic rocks may occur at temperatures significantly lower than those given by most other investigators (~ 85 to $115°$ C). Extending their investigation to δD-determinations, they showed that meteoric groundwater is important in the serpentinization of many ultramafic rocks. Oceanic ultramafic samples show a uniform isotopic composition, suggesting that seawater was involved in the serpentinization process.

3. D/H Ratios in Metamorphic Rocks

TAYLOR and EPSTEIN (1966 b), and SHIEH and TAYLOR (1969 a) have studied hydrogen isotope variations during contact metamorphism. In their investigation they have analyzed muscovite, biotite, and hornblende and found without exception that muscovite concentrates deuterium relative to coexisting biotite. Hornblende and biotite have almost identical D/H ratios. When the hydrogen isotope data of minerals in contact metamorphic rocks and their associated plutons are compared with those from regional metamorphic rocks, it may be concluded that in general, contact metamorphic rocks tend to have lower D/H ratios than plutonic rocks (see Fig. 37). This implies that there was some type of interaction of H_2O (or H_2) between the aureole and the intrusive.

In the contact metamorphic rocks and associated plutons there is a rough correlation between D/H ratios and geographic locations of samples, which makes an isotope exchange with local meteoric groundwaters during contact metamorphism probable.

The D/H ratios of the pelitic rocks analyzed by SHIEH and TAYLOR (1969 a) do not show any obvious change across the contact metamorphic zones. The authors concluded that no significant fractionations of hydrogen isotopes accompany contact metamorphic dehydration reactions.

4. Carbon Isotope Ratios during Metamorphism

Marble is an abundant rock type in many regional metamorphic terrains. The calcite-dolomite assemblage is stable over a wide range of physical conditions. Carbon and oxygen isotope variations in meta-

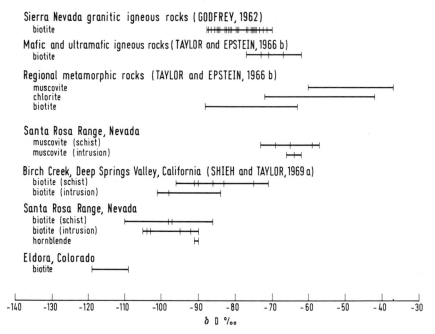

Fig. 37. Comparison of D/H ratios of minerals from contact metamorphic aureoles and their associated intrusions with D/H ratios of minerals from regional metamorphic and plutonic igneous rocks. (After SHIEH and TAYLOR, 1969a)

morphic dolomite-calcite assemblages have been studied by SCHWARCZ (1966), SHIEH and TAYLOR (1969b), and SHEPPARD and SCHWARCZ (1970).

$^{13}C/^{12}C$ and $^{18}O/^{16}O$ ratios of metamorphic carbonates span the range of diagenetically altered limestones. Metamorphic dolomites are isotopically heavier than calcites, an observation which has been interpreted as being due to an approach to isotopic equilibrium. The question of whether there is any systematic variation of isotopic composition with metamorphic grade cannot be answered at the moment. At least for the carbon isotope ratio, we may conclude that the $^{13}C/^{12}C$ ratio is not effected with increasing metamorphic grade.

There are numerous metamorphic decarbonation reactions, two of which are

$$\text{calcite} + \text{quartz} \rightleftharpoons \text{wollastonite} + CO_2$$
$$5\,\text{dolomite} + 8\,\text{quartz} + H_2O \rightleftharpoons 3\,\text{calcite} + \text{tremolite} + 7CO_2.$$

The CO_2 will escape out of the rock, and it is therefore interesting to inquire into the isotope effects that accompany such a process. If the liberated CO_2 is in isotopic equilibrium with the carbonate before escaping, the oxygen isotopic composition of the CO_2 can be estimated from the laboratory calibration curves (O'NEIL and EPSTEIN, 1966b). We know from these experiments that in the temperature range 500 to 700° C the CO_2 in equilibrium with dolomite would be about 5‰ higher in $^{18}O/^{16}O$ ratio than the dolomite. If the isotopic fractionations during decarbonation are kinetically controlled, the situation may be more complex. SHARMA and CLAYTON (1965) have measured the oxygen isotopic fractionation between CO_2 and dolomite during thermal decomposition of dolomite in vacuum. They showed that even during rapid decarbonation in which no attempt is made to attain equilibrium, the kinetic fractionations between CO_2 and dolomite in the temperature range studied are apparently similar to the equilibrium fractionations.

SHIEH and TAYLOR (1969b) demonstrated that calc-silicate skarns usually have the lowest $^{18}O/^{16}O$ and $^{13}C/^{12}C$ ratios in comparison with the adjacent rock units. They concluded that decarbonation reactions are largely responsible for the lowering of the $\delta^{18}O$ and $\delta^{13}C$ values. This lowering of the $^{18}O/^{16}O$ ratio in the rocks of the contact aureole is in contrast to the effects of contact metamorphic dehydration reactions, which generally produce no change in $^{18}O/^{16}O$ ratio. These differences are a result of the fact that CO_2-mineral oxygen isotope fractionations are significantly larger than H_2O-mineral fractionations and are also usually in the opposite direction.

Inasmuch as most marine limestones have $\delta^{13}C$-values close to zero, one would expect that on decarbonation the limestones would give off CO_2 with $\delta^{13}C$-values of about $+5.0‰$. However, most of the CO_2 sampled in geothermal areas (average about -3 to $-5‰$, see p.63) has a negative $\delta^{13}C$-value. This CO_2 can hardly be expected to be simply derived from the decarbonation of marine limestones.

Most graphites in metamorphic rocks gave results similar to the original organic material in the presursor sedimentary material (CRAIG, 1953; GAVELIN, 1957; LANDERGREN, 1961; HAHN-WEINHEIMER, 1960, 1966). No dependence of the $^{13}C/^{12}C$ ratios of graphites on increasing metamorphic grade has been detected, despite the theoretical prediction that the $^{13}C/^{12}C$ ratios of graphites should increase because of the splitting off of isotopically light hydrocarbons. However, it might be possible that the primary inhomogeneities are camouflaging this effect.

Graphites and carbonates frequently occur together in metamorphic rocks. $^{13}C/^{12}C$ data of both have been given by HAHN-WEINHEIMER (1966) and others. Carbon fractionation factors in the system graphite-calcite have been given by BOTTINGA (1969a). From the measured isotopic difference between calcite and graphite, we may tentatively calculate "isotope equilibrium temperatures", which in almost all cases show that isotope equilibrium cannot have been reached.

5. Sulfur Isotope Distribution in Metamorphic Rocks

Sulfur isotope data from metamorphic rocks (ore minerals from metamorphic ore deposits are excluded and not discussed here) are very scarce. SPEELMANN and SCHWARCZ (1966) have argued that with increasing metamorphic grade, a homogenization of sulfur isotope distribution takes place. However, unpublished results of SIEWERS have shown that this process seems to be more complex.

6. Variations of the $^{41}K/^{39}K$ Ratio in Contact Aureoles

The use of the $^{41}K/^{39}K$ ratio to detect and interpret movement of potassium in contact zones has been reported by SCHREINER and VERBEEK (1965), VERBEEK and SCHREINER (1967), and SCHREINER and WELKE (1971). The results are consistent with a diffusive transfer of potassium from the "hot" pluton into the "cold" sediments that would result in a preferential enrichment of the lighter isotope ^{39}K in the sediments.

X. Balance Calculations

Balance calculations of the average isotope ratio of the earth's crust are very problematical, because the numbers used are very rough estimations with a high degree of uncertainty. Balance computations of the following type are based on the assumption that all elements now found in the hydrosphere, in sediments, and in metamorphic and igneous crustal rocks must be derived from the earth's mantle.

As can be seen in Table 11, the mean calculated δ-values for hydrogen, carbon, and sulfur for the continental crust do not agree with experimental data, except for carbon. The discrepancies in the mean δ-values for hydrogen and sulfur cannot be explained in the same way. If we regard the oceans as the condensation product of water degassed from the earth during an early stage, then the mean δD-value of the oceans should represent the mean δD-value of the crust and the mantle. However, this is not the case.

In the upper zones of the atmosphere a photo-dissociation of water vapor into H_2 and O_2 takes place. Hydrogen, as the lightest gas, can continually escape from the gravity field of the earth. This process very probably proceeds parallel with an isotope fractionation causing a preferential loss of H and a retention of D. The magnitude of this effect is rather difficult to assess. However, one can say that there has probably been a progressive enrichment of deuterium in the ocean with time, due to the preferential escape of H relative to D.

Table 11. Isotopic balance calculation for the earth's crust for hydrogen, carbon, and sulfur

Reservoir	Mass in 10^{18} t	Average hydrogen con. in % H_2O	Mean isotopic δD-value in ‰	Average carbon conc. in % C	Mean isotopic δ-value in ‰	Average sulfur conc. in % S	Mean $\delta^{34}S$
Ocean water	1.4	—	0	0.003	0	0.1	+20
Sedimentary rocks	1.7	5%[a]	−50	2.4% carb. C[i] 0.5% org. C[i]	0 −28	0.74[a] as sulfate 0.25[a] as sulfide	+17[g] −15[h]
Igneous and metamorphic rocks	28.6	1%[b]	−70	0.03 in oxidized form[d] 0.02 in reduced form[d]	−10[e] −27[e]	0.04[a]	2.5[f]
Calculated mean δ-values			−13[c]		−7.5		7.2

[a] Estimated after RONOV and MIGDISOV (1971).
[b] Estimated after WEDEPOHL (1969).
[c] This value is even more uncertain than the other ones, since the amount of pore fluids has been neglected.
[d] After HOEFS (1965).
[e] After HOEFS (1972).
[f] Estimated after unpublished results of SIEWERS.
[g] After HOLSER and KAPLAN (1966).
[h] After KAPLAN et al. (1963).
[i] After RONOV (1968).

A mean calculated δD-value of $-13‰$ is therefore a rather meaningless number, except that it indicates that a process of preferential escape of H relative to D must take place.

The mean calculated δ^{13}C-value could be acceptable, because it coincides with the mean isotopic composition of carbonatites and diamonds. However, it must be emphasized that the amount of carbonate bound in sediments is very critical and also controversial.

The mean calculated δ^{34}S-value of $+7.2$ is certainly not correct. This could be due to an overestimation of the amount of evaporitic sulfate by RONOV and MIGDISOV (1971), who analyzed sediments only from the Russian platform. In this connection it must be noted that another attempt of a balance calculation by HOLSER and KAPLAN (1966) has given a mean value for the earth's crust of $6.0 \pm 2.3‰$. In both calculations the mean δ^{34}S-value is too heavy compared with potential mantle sulfur.

Many questions arise in this connection. Which naturally occurring specimens are representative for the mean δ-value of the earth's mantle? Are these the sulfides from basaltic dikes and sills with a mean δ-value of $0.1‰$ (SMITHERINGALE and JENSEN, 1963), which would be in relatively good agreement with the mean value of meteoritic material? Is this $1.3‰$, which SCHNEIDER (1970) gave for the alkali olivine basalts, showing, in his opinion, the least deviation from the original mantle value? Or does the average value lie between 2 and $3‰$, found by SIEWERS (unpublished data) for granitic rocks?

References

ABELSON, P. H., HOERING, T. C.: Carbon isotope fractionation in formation of amino acids by photosynthetic organisms. Proc. Nat. Acad. Sci. U.S. **47**, 623—632 (1961).

AGYEI, E. K., McMULLEN, C. C.: A study of the isotopic abundance of boron from various sources. Can. J. Earth Sci. **5**, 921—927 (1968).

ALDAZ, L., DEUTSCH, S.: On a relationship between air temperature and oxygen isotope ratio of snow and firn in the South Pole region. Earth Planet. Sci. Letters **3**, 267 (1967).

ALLENBY, R. J.: Determination of the isotopic ratios of silicon in rocks. Geochim. Cosmochim. Acta **5**, 40 (1954).

ANDERSON, A. T.: The dimensions of oxygen isotopic equilibrium attainment during prograde metamorphism. J. Geol. **75**, 323—332 (1967).

ANDERSON, A. T., CLAYTON, R. N., MAYEDA, T. K.: Oxygen isotope thermometry of mafic igneous rocks. J. Geol. **79**, 715—729 (1971).

ANGER, G., NIELSEN, H., PUCHELT, H., RICKE, W.: Sulfur isotopes in the Rammelsberg ore deposit (Germany). Econ. Geol. **61**, 511—536 (1966).

ARNASON, B.: The equilibrium constant for the fractionation of deuterium between ice and water. J. Phys. Chem. **73**, 3491 (1969).

AULT, W. A.: Isotopic fractionation of sulfur in geochemical processes. In: ABELSON, P. H. (Ed.): Researches in Geochemistry, p. 241. New York: John Wiley 1959.

AULT, W. A., KULP, J. L.: Isotopic geochemistry of sulfur. Geochim. Cosmochim. Acta **16**, 201—235 (1959).

AULT, W. A., KULP, J. L.: Sulfur isotopes and ore deposits. Econ. Geol. **55**, 73 (1960).

BACKUS, M., PINSON, W. H., HERZOG, L. F., HURLEY, P. M.: Calcium isotope ratios in the Homestead and Pasamonte meteorites and a Devonian limestone. Geochim. Cosmochim. Acta **28**, 735 (1964).

BACHINSKI, D. J.: Bond strength and sulfur isotope fractionation in coexisting sulfides. Econ. Geol. **64**, 56—65 (1969).

BAERTSCHI, P.: Messung und Deutung relativer Häufigkeitsvariationen von O^{18} und C^{13} in Karbonatgesteinen und Mineralen. Schweiz. Mineral. Petr. Mitt. **37**, 73—152 (1957).

BAINBRIDGE, K. T., NIER, A. O.: Relative isotopic abundances of the elements. Preliminary Rept. No. 9, Nuclear Science Series, Natl. Res. Council U.S., Washington D.C. (1950).

BALSIGER, H., GEISS, J., GROEGLER, N., WYTTENBACH, A.: Distribution and isotopic abundance of lithium in stone meteorites. Earth Planet. Sci. Letters **5**, 17—22 (1968).

BARNARD, G. P.: Modern mass spectrometry. London: Institute of Physics 1953.

BAUSCH, W., HOEFS, J.: Die Isotopenzusammensetzung von Dolomiten und Kalken aus dem süddeutschen Malm. Contr. Mineral. Petrol. **37**, 121 (1972).

BEGEMANN, F., FRIEDMAN, I.: Tritium and deuterium content of atmospheric hydrogen. Z. Naturforsch. **14a**, 1024—1031 (1959).

BELSKY, T., KAPLAN, I. R.: Light hydrocarbon gases, ^{13}C, and origin of organic matter in carbonaceous chondrites. Geochim. Cosmochim. Acta **34**, 257—278 (1970).

BERTENRATH, R., FRIEDRICHSEN, H., HELLNER, E.: Die Fraktionierung der Sauerstoffisotope ^{18}O/^{16}O im System Eisenoxid-Wasser. Deutsche Mineralogische Gesellschaft Sektion Geochemie Frühjahrstagung (1972) (Abstract).

BIGELEISEN, J.: Chemistry of isotopes. Science **147**, 463—471 (1965).

BIGELEISEN, J., MAYER, M. G.: Calculation of equilibrium constants for isotopic exchange reactions. J. Chem. Phys. **15**, 261 (1947).

BOATO, G.: The isotopic composition of hydrogen and carbon in the carbonaceous chondrites. Geochim. Cosmochim. Acta **6**, 209 (1954).

BOATO, G.: Meaning of deuterium abundance in meteorites. Nature **177**, 424 (1956).

BOIGK, H., STAHL, W.: Zum Problem der Entstehung nordwestdeutscher Erdgaslagerstätten. Erdöl u. Kohle **23**, 325 (1970).

BOKHOVEN, D., THEEUWEN, H. H. J.: Deuterium content of some natural organic substances. Koninkl. Ned. Akad. Wetenschap. Ser. **B 59**, 78 (1956).

BOTH, R. A., RAFTER, T. A., SOLOMON, M., JENSEN, M. L.: Sulfur isotopes and zoning of the Zeehan Mineral Field, Tasmania. Econ. Geol. **64**, 618—628 (1969).

BOYD, A. W., BROWN, F., LOUNSBURY, M.: Mass spectrometric study of natural and neutron-irradiated chlorine. Can. J. Phys. **33**, 35 (1955).

BOTTINGA, Y.: Calculated fractionation factors for carbon and hydrogen isotope exchange in the system calcite-carbondioxide-graphite-methane-hydrogen-water vapor. Geochim. Cosmochim. Acta **33**, 49—64 (1969 a).

BOTTINGA, Y.: Carbon isotope fractionation between graphite, diamond, and carbon dioxide. Earth Planet. Sci. Letters **5**, 301—307 (1969 b).

BOWEN, N. L.: The evolution of igneous rocks. Princeton 1928.

BRODSKY, A. E.: Isotopenchemie. Berlin: Akademie-Verlag 1961.

BROECKER, W., OVERSBY, V. M.: Chemical equilibria in the Earth. New York: McGraw-Hill 1970.

BROWN, J. S.: Isotopic zoning of lead and sulfur in S.E. Missouri. In: Genesis of stratiform Pb, Zn, Ba, F deposits. Econ Geol. Monograph **3**, 410 (1967).

BRUNÉE, C., VOSHAGE, H.: Massenspektrometrie. München: K. Thiemig KG 1964.

BUCHS, A., KRUMMENACHER, D., NOETZLIN, J.: La composition isotopique du magnésium extrait de roches et de laves d'origine très différentes. Schweiz. Mineral. Petrog. Mitt. **45**, 847 (1965).

BUSCHENDORF, F., NIELSEN, H., PUCHELT, H., RICKE, W.: S-Isotopenuntersuchungen am Pyrit-Sphalerit-Baryt-Lager Meggen und an verschiedenen Devonevaporiten. Geochim. Cosmochim. Acta **27**, 501 (1963).

CAMERON, A. E.,, LIPPERT, E. L.: Isotopic composition of bromine in nature. Science **121**, 136 (1955).

CATANZARO, E. J., MURPHY, T. J.: Magnesium isotope ratios in natural samples. J. Geophys. Res. **71**, 1271 (1966).

CHENEY, E. S., JENSEN, M. L.: Stable carbon isotopic composition of biogenic carbonates. Geochim. Cosmochim. Acta **29**, 1331—1346 (1965).

CHENG, H. H., BREMNER, J. M., EDWARDS, A. P.: Variations of nitrogen-15 abundance in soils. Science **146**, 1574 (1964).

CLAYTON, R. N.: Oxygen isotope fractionation between calcium carbonate and water. J. Chem. Phys. **34**, 724 (1961).

CLAYTON, R. N.: Oxygen isotope geochemistry: Thermometry of metamorphic rocks. In: SHAW, D. (Ed.): Studies in Analytical Geochemistry, p. 42. Toronto: University of Toronto Press 1963 a.

CLAYTON, R. N.: Carbon isotope abundance in meteoritic carbonates. Science **140**, 192 (1963 b).

CLAYTON, R. N., EPSTEIN, S.: The relationship between O^{18}/O^{16} ratios in coexisting quartz, carbonate and iron oxides from various geological deposits. J. Geol. **66**, 352—371 (1958).

CLAYTON, R. N., DEGENS, E. T.: Use of carbon isotope analyses of carbonates for differentiation of fresh-water and marine sediments. Bull. Am. Ass. Petrol. Geol. **43**, 890 (1959).

CLAYTON, R. N., MAYEDA, T. K.: The use of bromine pentafluoride in the extraction of oxygen from oxides and silicates for isotopic analysis. Geochim. Cosmochim. Acta **27**, 43—52 (1963).

CLAYTON, R. N., FRIEDMAN, I., GRAF, D. L., MAYEDA, T. K., MEENTS, W. F., SHIMP, N. F.: The origin of saline formation waters. 1. Isotopic composition. J. Geophys. Res. **71**, 3869—3882 (1966).

CLAYTON, R. N., JONES, B. F., BERNER, R. A.: Isotope studies of dolomite formation under sedimentary conditions. Geochim. Cosmochim. Acta **32**, 415 (1968 a).

CLAYTON, R. N., SKINNER, H. C. W., BERNER, R. A., RUBINSON, M.: Isotopic compositions of recent South Australian lagoonal carbonates. Geochim. Cosmochim. Acta **32**, 983 (1968 b).

CLAYTON, R. N., MUFFLER, L. J. P., WHITE, D. E.: Oxygen isotope study of calcite and silicates of the River Ranch No. 1 well, Salton Sea Geothermal Field, California. Am. J. Sci. **266**, 968 (1968 c).

CLAYTON, R. N., ONUMA, N., MAYEDA, T. K.: Oxygen isotope fractionation in Apollo 12 rocks and soils. Proc. Second Lunar Sci. Conf. Vol. 2, pp. 1417—1420. Massachusetts: M.I.T. Press 1971.

CLAYTON, R. N., O'NEIL, J. R., MAYEDA, T. K.: Oxygen isotope exchange between quartz and water. J. Geophys. Res. **77**, 3057 (1972).

COLOMBO, U., GAZZARRINI, F., GONFIANTINI, R., SIRONI, G., TONGIORGI, E.: Measurements of C^{13}/C^{12} isotope ratios on Italian natural gases and their geochemical interpretation. Advances in Organic Geochemistry 1964, pp. 279—293. Oxford: Pergamon Press 1966.

COLOMBO, U., GAZZARRINI, F., GONFIANTINI, R., KNEUPER, G., TEICHMÜLLER, M., TEICHMÜLLER, R.: Das C^{12}/C^{13}-Verhältnis von Kohlen und kohlenbürtigem Methan. Compte Rendu 6e Congrès Intern. Strat. Géol. Carbonif., Sheffield 1967, Volume II — 1970 —, p. 557—574.

COMPSTON, W.: The carbon isotopic composition of certain marine invertebrates and coal from the Australian Permian. Geochim. Cosmochim. Acta **18**, 1 (1960).

COMPSTON, W., EPSTEIN, S.: A method for the preparation of carbon dioxide from water vapor for oxygen isotope analysis. Trans. Am. Geophys. Union **39**, 511 (1958).

CONWAY, C. M., TAYLOR, H. P.: $^{18}O/^{16}O$ and $^{13}C/^{12}C$ ratios of coexisting minerals in the Oka and Magnet Cove carbonatite bodies. J. Geol. **77**, 618 (1969).

CORLESS, J. T.: Observations on the isotopic geochemistry of calcium. Earth Planet. Sci. Letters **4**, 475 (1968).

CORTECCI, G., LONGINELLI, A.: Isotopic composition of sulfate in rain water, Pisa, Italy. Earth Planet. Sci. Letters **8**, 36 (1970).

CRAIG, H.: The geochemistry of the stable carbon isotopes. Geochim. Cosmochim. Acta **3**, 53 (1953).

CRAIG, H.: Isotopic standards for carbon and oxygen and correction factors for mass-spectrometric analysis of carbon dioxide. Geochim. Cosmochim. Acta **12**, 133 (1957).

CRAIG, H.: Isotopic variations in meteoric waters. Science **133**, 1702 (1961 a).

CRAIG, H.: Standard for reporting concentrations of deuterium and oxygen-18 in natural waters. Science **133**, 1833 (1961 b).

CRAIG, H.: The isotopic geochemistry of water and carbon in geothermal areas. Proc. Spoleto Conference on Nuclear Geology, Ed.: E. TONGIORGI, Spoleto 1963.

CRAIG, H.: The measurement of oxygen isotope paleotemperatures. Proc. Spoleto Conference on Stable Isotopes in Oceanographic Studies and Paleotemperatures **3** (1965).

CRAIG, H.: Isotopic composition and origin of the Red Sea and Salton Sea geothermal brines. Science **154**, 1544 (1966).

CRAIG, H.: Isotope separation by carrier diffusion. Science **159**, 93—96 (1967).

CRAIG, H., KEELING, C. D.: Effects of atmospheric N_2O on the measured isotopic composition of atmospheric CO_2. Geochim. Cosmochim. Acta **27**, 549 (1963).

CRAIG, H., BOATO, G., WHITE, D. E.: Isotopic geochemistry of thermal waters. Proc. 2nd Conf. on Nuclear Processes in Geologic settings (1956), p. 29.

CRAIG, H., GORDON, L., HORIBE, Y.: Isotopic exchange effects in the evaporation of water. J. Geophys. Res. **68**, 5079 (1963).

CRAIG, H., GORDON, L.: Deuterium and oxygen-18 variations in the ocean and the marine atmosphere. In: Symposium on marine geochemistry. Graduate School of Oceanography, University of Rhode Island, Occ. Publ. No. 3, 277 (1965).

DANSGAARD, W.: Stable isotopes in precipitation. Tellus **16**, 436—468 (1964).

DANSGAARD, W., TAUBER, H.: Glacier oxygen-18 content and Pleistocene ocean temperatures. Science **166**, 499—502 (1969).

DANSGAARD, W., JOHNSEN, S. J., MOLLER, J., LANGWAY, C. C.: One thousand centuries of climatic record from Camp Century on the Greenland ice sheet. Science **166**, 377—381 (1969).

DAVIS, J. B., KIRKLAND, D. W.: Native sulfur deposition in the Castile formation, Gulberson County, Texas. Econ. Geol. **65**, 107 (1970).

DAUGHTRY, A. C., PERRY, D., WILLIAMS, M.: Magnesium isotope distribution in dolomite. Geochim. Cosmochim. Acta **26**, 857 (1962).

DECHOW, E.: Geology, sulfur isotopes, and the origin of the Heath Steele ore deposits, Newcastle, N.B., Canada. Econ. Geol. **55**, 539—556 (1960).

DECHOW, E., JENSEN, M. L.: Sulfur isotopes of some Central African sulfide deposits. Econ. Geol. **60**, 894—941 (1965).

DEEVEY, E. S., STUIVER, M.: Distribution of natural isotopes of carbon in Linsley Pond and other New England lakes. Limnol. Oceanog. **9**, 1 (1964).

DEGENS, E. T.: Biogeochemistry of stable carbon isotopes. In: G. EGLINTON, M. T. J. MURPHY (Eds.): Organic Geochemistry. Berlin-Heidelberg-New York: Springer 1969.

DEGENS, E. T., EPSTEIN, S.: Oxygen and carbon isotope ratios in coexisting calcites and dolomites from recent and ancient sediments. Geochim. Cosmochim. Acta **28**, 23 (1964).

DEGENS, E. T., GUILLARD, R. R. L., SACKETT, W. M., HELLEBUST, J. A.: Metabolic fractionation of carbon isotopes in marine plankton. I. Temperature and respiration experiments. Deep-Sea Res. **15**, 1—9 (1968a).

DEGENS, E. T., BEHRENDT, M., GOTTHARDT, G., REPPMANN, E.: Metabolic fractionation of carbon isotopes in marine plankton. II. Data on samples collected off the coasts of Peru and Ecuador. Deep-Sea Res. **15**, 11—20 (1968b).

DEINES, P.: The carbon and oxygen isotopic composition of carbonates from the Oka carbonatite complex, Quebec, Canada. Geochim. Cosmochim. Acta **34**, 1199 (1970).

DESSAU, G., JENSEN, M. L., NAKAI, N.: Geology and isotopic studies of Silician sulfur deposits. Econ. Geol. **57**, 410—438 (1962).

DEUSER, W. G.: Extreme $^{13}C/^{12}C$ variations in Quaternary dolomites from the continental shelf. Earth. Planet. Sci. Letters **8**, 118 (1970).

DEUSER, W. G., DEGENS, E. T.: Carbon isotope fractionation in the system CO_2 (gas)-CO_2(aqueous)-HCO_3(aqueous). Nature **215**, 1033 (1967).

DEUSER, W. G., HUNT, J. M.: Stable isotope ratios of dissolved inorganic carbon in the Atlantic. Deep-Sea Res. **16**, 221—225 (1969).

DEUTSCH, S., AMBACH, W., EISNER, H.: Oxygen isotope study of snow and firn on an Alpine glacier. Earth Planet. Sci. Letters **1**, 197 (1966).

DEVEREUX, I.: Oxygen isotope ratios of minerals from the regionally metamorphosed schists of Otago, New Zealand. N.Z.J. Sci. **11**, 526 (1968).

DEWS, J. R.: The isotopic composition of lithium in chondrules. J. Geophys. Res. **71**, 4011 (1966).

DOLE, M., JENKS, G.: Isotopic composition of photosynthetic oxygen. Science **100**, 409 (1944).

DOLE, M., LANE, G. A., RUDD, D. P., ZAUKELIES, D. A.: Isotopic composition of atmospheric oxygen and nitrogen. Geochim. Cosmochim. Acta **6**, 65 (1954).

DUCKWORTH, H. E.: Mass-spectroscopy. Cambridge At the University Press 1958.

DUPLESSY, J. C., LALOU, C., GOMES DE AZEVEDO, A. E.: Concretioning conditions in cave studies with stable isotopes of oxygen and carbon. Compt. Rend. **268**, 2327 (1969).

ECKELMANN, W. R., BROECKER, W. S., WHITLOCK, D. W., ALLSUP, J. R.: Implications of carbon isotopic composition of total organic carbon of some recent sediments and ancient oils. Bull. Am. Ass. Petrol. Geol. **46**, 699 (1962).

EDWARDS, G.: Isotopic composition of meteoritic hydrogen. Nature **176**, 109 (1955).

EHHALT, D., KNOTT, K., NAGEL, J. F., VOGEL, J. C.: Deuterium and oxygen-18 in rain water. J. Geophys. Res. **68**, 3775 (1963).

EICHLER, R., RISTEDT, H.: Isotopic evidence on the early life history of Nautilus pumpilius (Linné). Science **153**, 734—736 (1966).

EICHMANN, R.: Das Stickstoffisotopenverhältnis und die Zusammensetzung einiger nordwestdeutscher Erdgase und Erdölgase. Diss. Universität Braunschweig 1969.

EMILIANI, C.: Pleistocene temperatures. J. Geol. **63**, 538 (1955).

EMILIANI, C.: Paleotemperature analysis of core 280 and Pleistocene correlations. J. Geol. **66**, 264 (1958).

EMILIANI, C.: Isotopic paleotemperatures. Science **154**, 851 (1966).

EMRICH, K., EHHALT, D. H., VOGEL, J. C.: Carbon isotope fractionation during the precipitation of calcium carbonate. Earth Planet. Sci. Letters **8**, 363—371 (1970).

ENGEL, A. E., CLAYTON, R. N., EPSTEIN, S.: Variations in the isotopic composition of oxygen and carbon in Leadville limestone in its hydrothermal and metamorphic phases. J. Geol. **66**, 374 (1958).

EPSTEIN, S.: The variations of the O-18/O-16 ratio in nature and some geological applications. In: ABELSON, P. H. (Ed.): Researches in Geochemistry, p. 217. New York-London: John Wiley and Sons 1959.

EPSTEIN, S., BUCHSBAUM, H. A., LOWENSTAM, H. A., UREY, H. C.: Revised carbonate-water isotopic temperature scale. Bull. Geol. Soc. Am. **64**, 1315 (1953).

EPSTEIN, S., MAYEDA, T. K.: Variations of the O^{18}/O^{16} ratio in natural waters. Geochim. Cosmochim. Acta **4**, 213 (1953).

EPSTEIN, S., LOWENSTAM, H. A.: Temperature-shell-growth relations of recent and interglacial Pleistocene shoal-water biota from Bermuda. J. Geol. **61**, 424—438 (1953).

EPSTEIN, S., SHARP, R. P.: Oxygen isotope variations in the Malaspina and Saskatchewan glaciers. J. Geol. **67**, 88 (1959).

EPSTEIN, S., GRAF, D. L., DEGENS, E. T.: Oxygen isotope studies on the origin of dolomites. In: Isotope and cosmic chemistry, p. 169. Amsterdam: North Holland Publ. Comp. 1964.

EPSTEIN, S., SHARP, R. P., GOW, A. J.: Six-year record of oxygen and hydrogen isotope variations in South Pole firn. J. Geophys. Res. **70**, 1809 (1965).

EPSTEIN, S., SHARP, R. P., GOW, A. J.: Antarctic ice sheet: Stable isotope analyses of Byrd station cores and interhemispheric climatic implications. Science **168**, 1570—1572 (1970).

EPSTEIN, S., TAYLOR, H. P.: Variations of O-18/O-16 in minerals and rocks. In: ABELSON, P. H. (Ed.): Researches in Geochemistry, Vol. **2**, 29. New York-London: John Wiley and Sons 1967.

EPSTEIN, S., TAYLOR, H. P.: $^{18}O/^{16}O$, $^{30}Si/^{28}Si$, D/H and $^{13}C/^{12}C$ studies of lunar rocks and minerals. Science **167**, 533—536 (1970).

EPSTEIN, S., TAYLOR, H. P.: O^{18}/O^{16}, Si^{30}/Si^{28}, D/H and C^{13}/C^{12} ratios in lunar samples. Proc. Second Lunar Sci. Conf. **2**, 1421—1441, The M.I.T. Press 1971.

ESLINGER, E. V., SAVIN, S. M.: Oxygen isotope studies of burial metamorphism of the Belt Supergroup, Glacier National Park, Montana. GSA meeting 1971, Program p. 558.

EWALD, H., HINTENBERGER, H.: Methoden und Anwendungen der Massenspektroskopie. Weinheim: Verlag Chemie 1953.

FEELY, H. W., KULP, L.: Origin of Gulf coast salt-dome sulphur deposits. Bull. Am. Ass. Petr. Geologists **41**, 1802 (1957).

FINLEY, H. O., EBERLE, A. R., RODDEN, C. J.: Isotopic composition of certain boron minerals. Geochim. Cosmochim. Acta **26**, 911 (1962).

FISCHER, W., MAASS, I., SONTAG, E., SÜSS, M.: $^{12}C/^{13}C$-Untersuchungen an petrologisch, chemisch und technologisch charakterisierten Braunkohlenproben. Z. Angew. Geol. **14**, 182 (1968).

FISCHER, W., MAASS, I., SONTAG, E.: $^{12}C/^{13}C$-Untersuchungen an Braunkohleinhaltsstoffen. Z. Angew. Geol. **16**, 126 (1970).

FONTES, J. C., GONFIANTINI, R.: Fractionnement isotopique de l'hydrogène dans l'eau de cristallisation du gypse. Acad. Sci. Compt. Rend. **2650**, 4—6 (1967).

FONTES, J. C., FRITZ, P., LETOLLE, R.: Composition isotopique, minéralogique et genèse des dolomies du Bassin de Paris. Geochim. Cosmochim. Acta **34**, 279—294 (1970).

FORNACA-RINALDI, G., PANICHI, C., TONGIORGI, E.: Some causes of the isotopic composition of carbon and oxygen in cave concretions. Earth Planet. Sci. Letters **4**, 321 (1968).

FRANK, D. J., SACKETT, W. M.: Kinetic isotope effects in the thermal cracking of neopentane. Geochim. Cosmochim. Acta **33**, 811—820 (1969).

FRIEDMAN, G. M.: Early diagenesis and lithification in carbonate sediments. J. Sediment. Petrol. **34**, 777—813 (1964).

FRIEDMAN, I.: Deuterium content of natural waters. Geochim. Cosmochim. Acta **4**, 89 (1953).

FRIEDMAN, I.: Water in tektites. Proc. 2nd Conf. on Nuclear Processes in Geologic Settings (1955), p. 1.

FRIEDMAN, I., SMITH, R. L.: The deuterium content of water in some volcanic glasses. Geochim. Cosmochim. Acta **15**, 218—228 (1958).

FRIEDMAN, I., MACHTA, L., SOLLER, R.: Water vapor exchange between a water droplet and its environment. J. Geophys. Res. **67**, 2761—2766 (1962).

FRIEDMAN, I., HALL, W. A.: Fractionation of O^{18}/O^{16} between coexisting calcite and dolomite. J. Geol. **71**, 238—243 (1963).

FRIEDMAN, I., REDFIELD, A. C., SCHOEN, B., HARRIS, J.: The variation in the deuterium content of natural waters in the hydrologic cycle. Rev. Geophys. **2**, 177—224 (1964).

FRIEDMAN, I., O'NEIL, J. R., ADAMI, L. H., GLEASON, J. D., HARDCASTLE, K.: Water, hydrogen, deuterium, carbon, carbon-13, and oxygen-18 content of selected lunar material. Science **167**, 538—540 (1970).

FRIEDMAN, L., IRSA, A. P.: Variations in the isotopic composition of carbon in urban atmospheric carbon dioxide. Science **158**, 263—264 (1967).

FRIEDRICH, G., SCHACHNER, D., NIELSEN, H.: S-Isotopenuntersuchungen an Sulfiden aus dem Erzvorkommen der Sierra de Cartagena in Spanien. Geochim. Cosmochim. Acta **28**, 683 (1964).

FRIEDRICHSEN, H.: Oxygen isotope fractionation between coexisting minerals of the Grimstad granite. Neues Jahrb. Mineral. Monatsh. **1**, 26 (1971).

FRITZ, P.: The oxygen and carbon isotopic composition of carbonates from the Pine Point lead-zinc ore deposits. Econ. Geol. **64**, 733 (1969).

FRITZ, P., SMITH, D. G. W.: The isotopic composition of secondary dolomites. Geochim. Cosmochim. Acta **34**, 1161 (1970).

FRITZ, P.: Geochemical characteristics of dolomites and the ^{18}O-content of Middle Devonian oceans. Earth Planet. Sci. Letters **11**, 277 (1971).

GALIMOV, E. M., GRINENKO, U. A.: The effect of leaching under surface conditions on the isotopic composition of carbon in secondary calcite. Geochemistry **1**, 79—82 (1965).

GARLICK, G. D.: Oxygen isotope fractionation in igneous rocks. Earth Planet. Sci. Letters **1**, 361 (1966).

GARLICK, G. D.: The stable isotopes of oxygen. In: WEDEPOHL, K. H. (Ed.): Handbook of Geochemistry, 8B. Berlin-Heidelberg-New York: Springer 1969.

GARLICK, G. D., EPSTEIN, S.: The isotopic composition of oxygen and carbon in hydrothermal minerals at Butte, Montana. Econ. Geol. **61**, 1325 (1966).

GARLICK, G. D., EPSTEIN, S.: Oxygen isotope ratios in coexisting minerals of regionally metamorphosed rocks. Geochim. Cosmochim. Acta **31**, 181 (1967).

GAVELIN, S.: Variations in isotopic composition of carbon from metamorphic rocks in Northern Sweden and their geological significance. Geochim. Cosmochim. Acta **12**, 297—314 (1957).

GAVELIN, S., PARVEL, A., RYHAGE, R.: Sulfur isotope fractionation in sulfur mineralisation. Econ. Geol. **55**, 510—530 (1960).

GEHLEN, K. VON: Schwefel-Isotope und die Genese von Erzlagerstätten. Geol. Rundschau **55**, 178—196 (1965).

GEHLEN, K. VON, NIELSEN, H., RICKE, W.: S-Isotopen-Verhältnisse in Baryt und Sulfiden aus hydrothermalen Gängen im Schwarzwald und jüngeren Barytgängen in Süddeutschland und ihre genetische Bedeutung. Geochim. Cosmochim. Acta **26**, 1189—1207 (1962).

GODFREY, J. D.: The deuterium content of hydrous minerals from the East-Central Sierra Nevada and Yosemite National Park. Geochim. Cosmochim. Acta **26**, 1215—1245 (1962).

GONFIANTINI, R.: Some results on oxygen isotope stratigraphy in the deep drilling at King Baudouin Station, Antarctica. J. Geophys. Res. **70**, 1815 (1965).

GOW, A. J., EPSTEIN, S.: On the use of stable isotopes to trace the origins of ice in a floating ice tongue. J. Geophys. Res. **77**, 6552—6557 (1972).

GRADSZTAJN, E., SALOMÉ, M., YANIV, A., BERNAS, R.: Isotopic analysis of lithium in the Holbrook meteorite and in terrestrial samples with a sputtering ion source mass spectrometer. Earth Planet. Sci. Letters **3**, 387 (1968).

GREIG, J. A., BAADSGAARD, H., CUMMING, G. L., FOLINSBEE, R. E., KROUSE, H. R., OHMOTO, H., SASAKI, A., SMEJKAL, V.: Lead and sulphur isotopes of the Irish base metal mines in Carboniferous carbonate host rocks. Soc. Mining Geol. Japan, Spec. Issue **2**, 84—92 (1971).

GROOTENBOER, J., SCHWARCZ, H. P.: Experimentally determined sulfur isotope fractionations between sulfide minerals. Earth Planet. Sci. Letters **7**, 162—166 (1969).

GROSS, M. G.: Variations in the O^{18}/O^{16} and C^{13}/C^{12} ratios of diagenetically altered limestones in the Bermuda Islands. J. Geol. **72**, 170 (1964).

GROSS, M. G., TRACEY, J. I.: Oxygen and carbon isotopic composition of limestones and dolomites, Bikini and Eniwetok Atolls. Science **151**, 1082 (1966).

GROSS, W. H., THODE, H. G.: Ore and the source of acid intrusives using sulfur isotopes. Econ. Geol. **60**, 576—589 (1965).

HAHN-WEINHEIMER, P.: Boron and carbon content of basic and intermediate metamorphites of the Munchberg gneiss massif and their C^{12}/C^{13} ratios. Inter. Geol. Congr., Rept 21st Session, Copenhagen 1960, part **13**, 431—442 (1960).

HAHN-WEINHEIMER, P.: Die isotopische Verteilung von Kohlenstoff und Schwefel in Marmor und anderen Metamorphiten. Geol. Rundschau **55**, 197—208 (1966).

HALL, W. E., FRIEDMAN, I.: Oxygen and carbon isotopic composition of ore and host rock of selected Mississippi Valley deposits. U.S. Geol. Survey Prof. Paper **650-C**, C-140 to C-148 (1969).

HARRISON, A. G., THODE, H. G.: The kinetic isotope effect in the chemical reduction of sulphate. Faraday Soc. Trans. **53**, 1648—1651 (1957a).

HARRISON, A. G., THODE, H. G.: Mechanism of the bacterial reduction of sulphate from isotope fractionation studies. Faraday Soc. Trans. **54**, 84—92 (1957b).

HARRISON, A. G., THODE, H. G.: Sulphur isotope abundances in hydrocarbons and source rocks of Uinta Basin, Utah. Bull. Am. Ass. Petrol. Geologists **42**, 2642 (1958).

HARTECK, P., SUESS, H.: Der Deuteriumgehalt des freien Wasserstoffs in der Erdatmosphäre. Naturwissenschaften **36**, 218 (1949).

HARTMANN, M., NIELSEN, H.: δ^{34}S-Werte in rezenten Meeressedimenten und ihre Deutung am Beispiel einiger Sedimentprofile aus der westlichen Ostsee. Geol. Rundschau **58**, 621—655 (1969).

HENDY, C. H.: The isotopic geochemistry of speleothems. I. The calculation of the effects of different modes of formation on the isotopic composition of speleothems and their applicability as palaeoclimatic indicators. Geochim. Cosmochim. Acta **35**, 801 (1971).

HENDY, C. H., WILSON, A. T.: Palaeoclimatic data from speleothems. Nature **216**, 48 (1968).

HINTENBERGER, H.: Precision determinations of isotope abundances. In: Advan. mass spectrometry **3**, 517 (1966).

HIRT, B., EPSTEIN, S.: Die Isotopenzusammensetzung von natürlichem Calcium. Helv. Phys. Acta **37**, 179 (1964).

HODGSON, W. A.: Carbon and oxygen isotope ratios in diagenetic carbonates from marine sediments. Geochim. Cosmochim. Acta **30**, 1223—1233 (1966).

HOEFS, J.: Ein Beitrag zur Geochemie des Kohlenstoffs in magmatischen und metamorphen Gesteinen. Geochim. Cosmochim. Acta **29**, 399 (1965).

HOEFS, J.: Kohlenstoff- und Sauerstoff-Isotopenuntersuchungen an Karbonatkon-
kretionen und umgebendem Gestein. Contr. Mineral. Petrol. **27**, 66—79 (1970).

HOEFS, J.: Carbon isotope composition of igneous rocks. (unpublished manuscript)
(1972).

HOEFS, J., SCHIDLOWSKI, M.: Carbon isotope composition of carbonaceous matter
from the Precambrian of the Witwatersrand system. Science **155**, 1096 (1967).

HOEFS, J., GRAESER, S.: Schwefelisotopenuntersuchungen an Sulfiden und Sulfosal-
zen des Binnatales. Contr. Mineral. Petrol. **17**, 165—172 (1968).

HOEFS, J., EPSTEIN, S.: O^{18}/O^{16} ratios of minerals from migmatites, Rapakivi granites and orbicular rocks. Lithos **2**, 1—8 (1969).

HOERING, T.: Variations in nitrogen-15 abundance in naturally occurring substances. Science **122**, 1233 (1955).

HOERING, T.: Variations in the nitrogen isotope abundance. Proc. 2nd Conf. Nucl.
Processes in Geologic Settings, p. 85 (1956).

HOERING, T.: The isotopic composition of the ammonia and the nitrate ion in rain
water. Geochim. Cosmochim. Acta **12**, 97 (1957).

HOERING, T.: The carbon isotope effect in the synthesis of diamond. Carnegie Inst.
Wash. Geophys. Lab. Yearbook 1960—61, 204 (1961).

HOERING, T., MOORE, H. E.: The isotopic composition of the nitrogen in natural
gases and associated crude oils. Geochim. Cosmochim. Acta **13**, 225 (1958).

HOERING, T. C., PARKER, P. L.: The geochemistry of the stable isotopes of chlorine.
Geochim. Cosmochim. Acta **23**, 186 (1961).

HOLSER, W. T., KAPLAN, I. R.: Isotope geochemistry of sedimentary sulfates. Chem.
Geol. **1**, 93—135 (1966).

HULSTON, J. R., MCCABE, W. J.: Mass spectrometer measurements in the thermal
areas of New Zealand. Part II: Carbon isotopic ratios. Geochim. Cosmochim.
Acta **26**, 398—410 (1962).

HULSTON, J. R., THODE, H. G.: Variations in the S^{33}, S^{34} and S^{36} contents of meteorites and their relation to chemical and nuclear effects. J. Geophys. Res. **70**, 3475
(1965).

INGHRAM, M. G., HAYDEN, R. J.: A handbook on mass spectroscopy. Nuclear Science Series Report **14**, Nat. Acad. Sci.-Nat. Res. Council 1954.

JAMES, H. L., CLAYTON, R. N.: Oxygen isotope fractionation in metamorphosed iron
formations of the Lake Superior region, and in other iron-rich rocks. In: Petrologic studies (BUDDINGTON Volume). Geol. Soc. Am. **1962**, 217—239.

JENSEN, M. L.: Sulfur isotopes and mineral paragenesis. Econ. Geol. **52**, 269 (1957).

JENSEN, M. L.: Sulfur isotopes and the origin of sand-stone-type uranium deposits.
Econ. Geol. **53**, 598—616 (1958).

JENSEN, M. L.: Sulfur isotopes and hydrothermal mineral deposits. Econ. Geol. **54**,
374 (1959).

JENSEN, M. L.: Sulfur isotopes and mineral genesis. In: BARNES, H. L. (Ed.): Geochemistry of hydrothermal ore deposits. New York-London: Holt, Rinehart and
Winston Inc. 1967.

JENSEN, M. L., NAKAI, N.: Sources and isotopic composition of atmospheric sulfur.
Science **134**, 2102 (1961).

JENSEN, M. L., NAKAI, N.: Sulfur isotope meteorite standards, results and recom-
mendations. In: Biogeochemistry of sulfur isotopes, NSF Symposium Volume,
ed. by M. L. JENSEN, p. 31—35 (1962).

JONES, G. E., STARKEY, R. L.: Fractionation of stable isotopes of sulfur by microorganisms and their role in deposition of native sulfur. Appl. Microbiol. **5**, 111
(1957).

KAJIWARA, Y.: Sulfur isotope study of the Kuroko-ores of the Sha Kanai No. 1 deposits, Akita Prefecture, Japan. Geochem. J. **4**, 157 (1971).

KAJIWARA, Y., KROUSE, H. R., SASAKI, A.: Experimental study of sulfur isotope fractionation between coexistent sulfide minerals. Earth Planet. Sci. Letters **7**, 271—277 (1969).

KAJIWARA, Y., KROUSE, H. R.: Sulfur isotope partitioning in metallic sulfide systems. Can. J. Earth Sci. **8**, 1397—1408 (1971).

KAPLAN, I. R., RAFTER, T. A., HULSTON, J. R.: Sulphur isotope variations in nature. VIII.: Application to some biogeochemical problems. N.Z.J. Sci. **3**, 338—361 (1960).

KAPLAN, I. R., EMERY, K. O., RITTENBERG, S. C.: The distribution and isotopic abundance of sulphur in recent marine sediments of Southern California. Geochim. Cosmochim. Acta **27**, 297—331 (1963).

KAPLAN, I. R., RITTENBERG, S. C.: Carbon isotope fractionation during metabolism of lactate by desulfovibrio desulfuricans. J. Gen. Microbiol. **34**, 213 (1964).

KAPLAN, I. R., HULSTON, J. R.: The isotopic abundance and content of sulfur in meteorites. Geochim. Cosmochim. Acta **30**, 479—496 (1966).

KAPLAN, I. R., SMITH, J. W.: Concentration and isotopic composition of carbon and sulfur in Apollo 11 lunar samples. Science **167**, 541 (1970).

KAPLAN, I. R., PETROWSKI, C.: Carbon and sulfur isotope studies on Apollo 12 lunar samples. Proc. Second Lunar Sci. Conf. Vol. **2**, 1397—1406. Massachusetts: M.I.T. Press 1971.

KATZ, J. J., CRESPI, H. L.: Deuterated organisms: cultivation and uses. Science **151**, 1187 (1966).

KEELING, C. D.: The concentration and isotopic abundance of carbon dioxide in rural areas. Geochim. Cosmochim. Acta **13**, 322 (1958).

KEELING, C. D.: The concentration and isotopic abundances of CO_2 in the atmosphere. Tellus **2**, 200—203 (1960).

KEELING, C. D.: The concentration and isotopic abundances of carbon dioxide in rural and marine air. Geochim. Cosmochim. Acta **24**, 277 (1961).

KEIL, K.: Meteorite composition. In: Handbook of Geochemistry I. Editor: K. H. WEDEPOHL. Berlin-Heidelberg-New York: Springer 1969.

KEITH, M. L., ANDERSON, G. M.: Radiocarbon dating — fictitious results with mollusk shells. Science **141**, 634—637 (1963).

KEITH, M. L., WEBER, J. N.: Isotopic composition and environmental classification of selected limestones and fossils. Geochim. Cosmochim. Acta **28**, 1787 (1964).

KEITH, M. L., ANDERSON, G. M., EICHLER, R.: Carbon and oxygen isotopic composition of mollusk shells from marine and fresh-water environment. Geochim. Cosmochim. Acta **28**, 1757 (1964).

KEMP, A. L. W., THODE, H. G.: The mechanism of the bacterial reduction of sulphate from isotopic fractionation studies. Geochim. Cosmochim. Acta **32**, 71—91 (1968).

KOKUBU, N., MAYEDA, T., UREY, H. C.: Deuterium content of minerals, rocks and liquid inclusion from rocks. Geochim. Cosmochim. Acta **21**, 247—256 (1961).

KRANKOWSKY, D., MÜLLER, O.: Isotopenhäufigkeit und Konzentration des Lithium in Steinmeteoriten. Geochim. Cosmochim. Acta **28**, 1625 (1964).

KRANKOWSKY, D., MÜLLER, O.: Isotopic composition and abundance of lithium in meteoritic matter. Geochim. Cosmochim. Acta **31**, 1833 (1967).

KRICHEVSKY, M. I., SESLER, F. D., FRIEDMAN, I., NEWELL, M.: Deuterium fractionation during molecular H_2 formation in a marine pseudomonad. J. Biol. Chem. **236**, 2520 (1961).

KROOPNICK, P., CRAIG, H.: Atmospheric oxygen: isotopic composition and solubility fractionation. Science **175**, 54 (1972).

KROUSE, H. R., THODE, H. G.: Thermodynamic properties and geochemistry of isotopic compounds of selenium. Can. J. Chem. **40**, 367—375 (1962).

KROUSE, H. R., MODZELESKI, V. E.: C^{13}/C^{12} abundances in components of carbonaceous chondrites and terrestrial samples. Geochim. Cosmochim. Acta **34**, 459—474 (1970).

KUKHARENKO, A. A., DONTSOVA, E. I.: A contribution to the problem of the genesis of carbonatites. Econ. Geol., USSR **1**, 31—46 (1964).

KULP, J. L., AULT, W. U., FEELY, H. W.: Sulphur isotope abundances in sulphide minerals. Econ. Geol. **51**, 139 (1956).

LANDERGREN, S.: The content of ^{13}C in the graphite-bearing magnetite ores and associated carbonate rocks in the Norberg mining district, Central Sweden. Geol. Foren. Stockholm Forh. **83**, 151 (1961).

LANE, G. A., DOLE, M.: Fractionation of oxygen isotopes during respiration. Science **123**, 574 (1956).

LANCELOT, J., SARAZIN, G., ALLEGRE, C. J.: Composition isotopique du plomb et du soufre des galènes liées aux formations sédimentaires. Interprétations géologiques et géophysiques. Contr. Mineral. Petrol. **32**, 315 (1971).

LAWRENCE, J. R., TAYLOR, H. P.: Utilization of δD and $\delta^{18}O$ of ancient kaolinites to determine δD and $\delta^{18}O$ of ancient meteoric waters. GSA meeting 1970, abstracts p. 605.

LAWRENCE, J. R., TAYLOR, H. P.: Deuterium and oxygen-correlation: Clay minerals and hydroxides in Quaternary soils compared to meteoric waters. Geochim. Cosmochim. Acta **35**, 993 (1971).

LEBEDEV, V. I., PROKOF'YEV, L. M., KIRILLOV, TARASOV, A. V.: Potassium isotope ratios in micas. Geochemistry **1966**, 1003.

LEHMANN, W. J., SHAPIRO, I.: Isotopic composition of boron and its atomic weight. Nature **183**, 1324 (1959).

LETOLLE, R.: Observations sur la composition isotopique du potassium de certaines roches. Compt. Rend. **254**, 2205 (1962).

LETOLLE, R.: Sur l'abondance relative de l'isotope-41 du potassium suivant son origine géologique. Compt. Rend. **257**, 3996 (1963).

LETOLLE, R.: Sur la composition isotopique du calcium des échantillons naturels. Earth Planet. Sci. Letters **5**, 207 (1968).

LLOYD, M. R.: Variations in the oxygen and carbon isotope ratios of Florida Bay mollusks and their environmental significance. J. Geol. **72**, 84 (1964).

LLOYD, R. M.: Oxygen-18 composition of oceanic sulfate. Science **156**, 56—59 (1967).

LLOYD, R. M.: Oxygen isotope behavior in the sulfate-water system. J. Geophys. Res. **73**, 6099—6110 (1968).

LONGINELLI, A.: Oxygen isotope composition of orthophosphate from shells of living marine organisms. Nature **207**, 716 (1965).

LONGINELLI, A.: Ratios of O-18/O-16 in phosphate and carbonate from living and fossil marine organisms. Nature **211**, 923 (1966).

LONGINELLI, A., TONGIORGI, E.: Oxygen isotopic composition of some right- and left-coiled foraminifera. Science **144**, 1004 (1964).

LONGINELLI, A., CRAIG, H.: Oxygen-18 variations in sulfate ions in sea-water and saline lakes. Science **146**, 56—59 (1967).

LONGINELLI, A., NUTI, S.: Oxygen-isotope ratios in phosphate from fossil marine organisms. Science **160**, 879—882 (1968a).

LONGINELLI, A., NUTI, S.: Oxygen isotopic composition of phosphorites from marine formations. Earth Planet. Sci. Letters **5**, 13—16 (1968 b).

LONGINELLI, A., CORTECCI, G.: Isotopic abundance of oxygen and sulfur in sulfate ions from river water. Earth Planet. Sci. Letters **7**, 376—380 (1970).

LORIUS, C., MERLIVAT, L., HAGEMANN, R.: Variation in the mean deuterium content of precipitations in Antarctica. J. Geophys. Res. **74**, 7027 (1969).

LOWENSTAM, H. A.: Mineralogy, O-18/O-16 ratios, and strontium and magnesium contents of recent and fossil brachiopods and their bearing on the history of the oceans. J. Geol. **69**, 241 (1961).

LUSK, J., CROCKET, J. H.: Sulfur isotope fractionation in coexisting sulfides from the Heath Steele B-1 Orebody, New Brunswick, Canada. Econ. Geol. **64**, 147—155 (1969).

MACNAMARA, J., THODE, H. G.: Comparison of the isotopic constitution of terrestrial and meteoritic sulphur. Phys. Rev. **78**, 307 (1950).

MATSUHISA, Y., HONMA, H., MATSUBAYA, O., SAKAI, H.: Oxygen isotopic study of the Cretaceous granitic rocks in Japan. Contr. Mineral. Petrol. **37**, 65—74 (1972).

MAUGER, R. L., LIVINGSTON, D. E., JENSEN, M. L., DAMON, P. E.: Sulfur and initial strontium isotopic investigations of basin and range intermediate rocks. Am. Geophys. Union, 48th Annual Meet., Program 254 (1967).

MAY, F., FREUND, W., MÜLLER, E. P., DOSTAL, K. P.: Modellversuche über Isotopenfraktionierung von Erdgaskomponenten während der Migration. Z. Angew. Geol. **14**, 376 (1968).

MAYNE, K. I.: Natural variations in the nitrogen isotope abundance ratio in igneous rocks. Geochim. Cosmochim. Acta **12**, 185 (1957).

MCCREA, J. M.: The isotopic chemistry of carbonates and a paleotemperature scale. J. Chem. Phys. **18**, 849 (1950).

MCDOWELL, C. A.: Mass Spectrometry. New York: McGraw Hill Book Co. 1963.

MCMULLEN, C. C., CRAGG, C. B., THODE, H. G.: Absolute ratio of $^{11}B/^{10}B$ in Searles Lake borax. Geochim. Cosmochim. Acta **23**, 147 (1961).

MEKHTIYEVA, V. L., PANKINA, G. R.: Isotopic composition of sulfur in aquatic plants and dissolved sulfates. Geochemistry **5**, 624—627 (1968).

MELANDER, L.: Isotope effects on reaction rates. New York: Ronald Press Co. 1960.

MELTON, C. E., GILPATRICK, L. O., BALDOCK, R., HEALY, R. M.: Improved techniques for the isotopic determination of boron on the mass spectrometer. Anal. Chem. **28**, 1049 (1956).

MIZUTANI, Y., RAFTER, T. A.: Oxygen isotopic composition of sulphates. III.: Oxygen isotopic fractionation in the bisulphate ion-water system. N.Z.J. Sci. **12**, 54—59 (1969 a).

MIZUTANI, Y., RAFTER, T. A.: Bacterial fractionation of oxygen isotopes in the reduction of sulphate and in the oxidation of sulphur. N.Z.J. Sci. **12**, 60—68 (1969 b).

MIZUTANI, Y., RAFTER, T. A.: Isotopic composition of sulphate in rain water, Gracefield, New Zealand. N.Z.J. Sci. **12**, 69—80 (1969 c).

MONSTER, J., ANDERS, E., THODE, H. G.: $^{34}S/^{32}S$ ratios for the different forms of sulphur in the Orgueil meteorite and their mode of formation. Geochim. Cosmochim. Acta **29**, 773—781 (1965).

MOORE, G. W.: Speleothem — A new cave term. Nat. Speleol. Soc. News **10** (6), 2 (1952).

MORTON, R. D., CATANZARO, E. J.: Stable chlorine isotope abundances in apatites from Odegardens Verk, Norway. Norsk Geol. Tidesskr. **44**, 307 (1964).

MÜLLER, G., NIELSEN, H., HOEFS, J.: Schwefel-Isotopen-Untersuchungen an Evaporiten der Kreuznacher Gruppe. Geol. Paläont. Monatsh. **12**, 745 (1966).

MÜNNICH, K. O., VOGEL, J. C.: Untersuchungen an pluvialen Wässern der Ost-Sahara. Geol. Rundschau **52**, 611 (1962).

MURATA, K. J., FRIEDMAN, I., MADSEN, B. M.: Carbon-13 rich diagenetic carbonates in Miocene formations of California and Oregon. Science **156**, 1484—1486 (1967).

MURPHEY, B. F., NIER, A. O.: Variations in the relative abundance of the carbon isotopes. Phys. Rev. **59**, 771 (1941).

NAKAI, N.: Carbon isotope fractionation of natural gas in Japan. J. Earth Sci. Nagoya Univ. **8**, 174 (1960).

NAKAI, N., JENSEN, M. L.: Biogeochemistry of sulfur isotopes. J. Earth Sci. Nagoya Univ. **8**, 181 (1960).

NAKAI, N., JENSEN, M. L.: The kinetic isotope effect in the bacterial reduction and oxidation of sulfur. Geochim. Cosmochim. Acta **28**, 1893—1912 (1964).

NAKAI, N., JENSEN, M. L.: Sources of atmospheric sulfur compounds. Geochemical J. **1**, 199—210 (1967).

NAUGHTON, J. J., TERADA, K.: Effect of eruption of Hawaiian volcanoes on the composition and carbon isotope content of associated volcanic and fumarolic gases. Science **120**, 580 (1954).

NIELSEN, H.: S-Isotope im marinen Kreislauf und das δ^{34}S der früheren Meere. Geol. Rundschau **55**, 160 (1965).

NIELSEN, H.: Schwefel-Isotopenverhältnisse aus St. Andreasberg und anderen Erzvorkommen des Harzes. N. Jb. Miner. Abh. (Rose Festband), **109**, 289 (1968a).

NIELSEN, H.: Data handling in precision isotope research. Advan. Mass Spectr. **4**, 267 (1968b).

NIELSEN, H.: Sulphur isotopes and the formation of evaporite deposits. In: Proc. Geology of saline deposits, Hannover (1968c).

NIELSEN, H.: Über den Fraktionierungsfaktor der bakteriellen Sulfatreduktion. Isotope titles (in press) (1972a).

NIELSEN, H.: Altersmäßige Zuordnung von Evaporiten und Solen auf Grund der δ^{34}S-Werte. Isotope titles (in press) (1972b).

NIELSEN, H., RICKE, W.: S-Isotopenverhältnisse von Evaporiten aus Deutschland. Ein Beitrag zur Kenntnis von δ^{34}S im Meerwasser-Sulfat. Geochim. Cosmochim. Acta **28**, 577 (1964).

NIER, A. O.: A redetermination of the relative abundances of the isotopes of carbon, nitrogen, oxygen, argon, and potassium. Phys. Rev. **77**, 789 (1950).

NIER, A. O., GULBRANSEN, E. A.: Variations in the relative abundance of the carbon isotopes. J. Am. Chem. Soc. **61**, 697 (1939).

NIER, A. O., NEY, E. P., INGHRAM, M. G.: A null method for the comparison of two ion currents in a mass spectrometer. Rev. Sci. Instr. **18**, 294 (1947).

NORTHROP, D. A., CLAYTON, R. N.: Oxygen isotope fractionations in systems containing dolomite. J. Geol. **74**, 174—196 (1966).

OANA, S., DEEVEY, E. S.: Carbon-13 in lake waters and its possible bearing on paleolimnology. Am. J. Sci. **258A**, 253—272 (1960).

OHMOTO, H.: Systematics of sulfur and carbon isotopes in hydrothermal ore deposits. Econ. Geol. **67**, 551—578 (1972).

OHMOTO, H., RYE, R. O.: The Bluebell Mine, British Columbia, I. Mineralogy, paragenesis, fluid inclusions and the isotopes of hydrogen, oxygen and carbon. Econ. Geol. **65**, 417—437 (1970).

O'NEIL, J. R.: Hydrogen and oxygen isotopic fractionation between ice and water. J. Phys. Chem. **72**, 3683 (1968).

O'NEIL, J. R., CLAYTON, R. N.: Oxygen isotope geothermometry. In: Isotope and cosmic chemistry. Amsterdam: North Holland Publ. Co. 1964.

O'NEIL, J. R., EPSTEIN, S.: A method for oxygen isotope analysis of milligram quantities of water and some of its applications. J. Geophys. Res. 71, 4955—4961 (1966a).

O'NEIL, J. R., EPSTEIN, S.: Oxygen isotope fractionation in the system dolomite-calcite-CO_2. Science 152, 198—201 (1966b).

O'NEIL, J. R., TAYLOR, H. P.: Oxygen isotope equilibrium between muscovite and water. Trans. Am. Geophys. Union 47, 212 (1966).

O'NEIL, J. R., TAYLOR, H. P.: The oxygen isotope and cation exchange chemistry of feldspars. Am. Mineralogist 52, 1414 (1967).

O'NEIL, J. R., HEDGE, C. E., JACKSON, E. D.: Isotopic investigations of xenoliths and host basalts from the Honolulu volcanic series. Earth Planet. Sci. Letters 8, 253 (1970).

ONUMA, N., CLAYTON, R. N., MAYEDA, T. K.: Oxygen isotope fractionation between minerals and an estimate of the temperature of formation. Science 167, 536—538 (1970).

ONUMA, N., CLAYTON, R. N., MAYEDA, T. K.: Oxygen isotope temperatures of "equilibrated" ordinary chondrites. Geochim. Cosmochim. Acta 36, 157 (1972a).

ONUMA, N., CLAYTON, R. N., MAYEDA, T. K.: Oxygen isotope cosmothermometer. Geochim. Cosmochim. Acta 36, 169 (1972b).

OSTLUND, C.: Isotopic composition of sulfur in precipitation and sea water. Tellus 11, 478—480 (1959).

OWEN, H. R., SCHAEFFER, O. A.: The isotopic abundance of chlorine from various sources. J. Am. Chem. Soc. 76, 2608 (1955).

PANKINA, R. G., MEKHIYEVA, V. L.: The isotopic composition of sulfur in the hydrogen sulfide of natural gas of the Bobrikovo Zone in the Volga-Urals Region. Geochemistry 1964, 853.

PARK, R., EPSTEIN, S.: Carbon isotope fractionation during photosynthesis. Geochim. Cosmochim. Acta 21, 110 (1960).

PARKER, P. L.: The biogeochemistry of the stable isotopes of carbon in a marine bay. Geochim. Cosmochim. Acta 28, 1155 (1964).

PARWEL, A., UBISCH, H. VON, WICKMAN, F. E.: On the variations in the relative abundance of boron isotopes in nature. Geochim. Cosmochim. Acta 10, 185 (1956).

PARWEL, A., RYHAGE, R., WICKMAN, F. E.: Natural variations in the relative abundances of the nitrogen isotopes. Geochim. Cosmochim. Acta 11, 165 (1957).

PERRY, E. C.: The oxygen isotope chemistry of ancient cherts. Earth Planet. Sci. Letters 3, 62 (1967).

PERRY, E. C., BONNICHSEN, B.: Quartz and magnetite: Oxygen-18/oxygen-16 fractionation in metamorphosed Biwabik Iron Formation. Science 153, 528 (1966).

PICCIOTTO, E., DE MAERE, X., FRIEDMAN, I.: Isotopic composition and temperature of formation of Antarctic snows. Nature 187, 857 (1960).

PILOT, J., RÖSLER, H. J., MÜLLER, P.: Zur geochemischen Entwicklung des Meerwassers und mariner Sedimente im Phanerozoikum mittels Untersuchungen von S-, O- und C-Isotopen. Neue Bergbautechnik 2, 161 (1972).

PINCKNEY, O. M., RAFTER, T. A.: Fractionation of sulfur isotopes during ore-deposition in the Upper Mississippi Valley Zinc-Lead District. Econ. Geol. 67, 315 (1972).

PINCKNEY, O. M., RYE, R. O.: Variation of O^{18}/O^{16}, C^{13}/C^{12}, texture and mineralogy in altered limestone in the Hill Mine, Cave-in-District. Illinois. Econ. Geol. **67**, 1—18 (1972).

PINUS, G. V.: Heavy oxygen isotope of the olivine from the ultramafic rocks as indicator of their genesis. J. Geophys. Res. **76**, 1339 (1971).

POSCHENRIEDER, W. P., HERZOG, R. F., BARRINGTON, A. E.: The relative abundance of the lithium isotopes in the Holbrook meteorite. Geochim. Cosmochim. Acta **29**, 1193 (1965).

PUCHELT, H., KULLERUD, G.: Sulfur isotope fractionation in the Pb-S system. Earth Planet. Sci. Letters **7**, 301 (1970).

PUCHELT, H., HOEFS, J., NIELSEN, H.: Sulfur isotope investigations of the Aegean volcanoes. Acta of the I. Intern. Sci. Congress on the Volcano of Thera, p. 303, 1971 a.

PUCHELT, H., SABELS, B. R., HOERING, T. C.: Preparation of sulfur hexafluoride for isotope geochemical analysis. Geochim. Cosmochim. Acta **35**, 625 (1971 b).

RABINOWITCH, E. I.: Photosynthesis and related processes, Vol. 1, p. 10. New York: Interscience 1945.

RAFTER, T. A.: Sulphur isotopic variations in nature. — Part 1 — The preparation of sulphur dioxide for mass spectrometer examination. N.Z.J. Sci. Tech. **B 38**, 849 (1957).

RAFTER, T. A.: Recent sulphur isotope measurements on a variety of specimens examined in New Zealand. Bull. Volcan. **28**, 3—20 (1965).

RAFTER, T. A., WILSON, S. H., SHILTON, B. W.: Sulphur isotopic variations in nature. Part 5: Sulphur isotopic variations in New Zealand geothermal borewaters. N.Z.J. Sci. **1**, 103 (1958 a).

RAFTER, T. A., WILSON, S. H., SHILTON, B. W.: Sulphur isotopic variations in nature. Part 6: Sulphur isotopic measurements on the discharge from fumaroles on White Island. N.Z.J. Sci. **1**, 154 (1958 b).

RAFTER, T. A., MIZUTANI, Y.: Oxygen isotopic composition of sulphates, part 2: Preliminary results on oxygen isotopic variation in sulphates and relationship to their environment and to their $\delta^{34}S$ values. N.Z.J. Sci. **10**, 816—840 (1967).

RANKAMA, K.: Isotope geology. London: Pergamon Press 1954.

RANKAMA, K.: Progress in isotope geology. New York: Interscience Publ. 1963.

REDFIELD, A. C., FRIEDMAN, I.: Factors affecting the distribution of deuterium in the ocean. In: Symposium on marine geochemistry, Univers. of Rhode Island, Occasional Publication **3**, 149—168 (1965).

REES, C. E.: The sulphur isotope balance of the ocean: an improved model. Earth Planet. Sci. Letters **7**, 366—370 (1970).

REUTER, J. H., EPSTEIN, S., TAYLOR, H. P.: O^{18}/O^{16} ratios of some chondritic meteorites and terrestrial ultramafic rocks. Geochim. Cosmochim. Acta **29**, 481—488 (1965).

REX, R. W., SYERS, J. K., JACKSON, M. L., CLAYTON, R. N.: Eolian origin of quartz in soils of Hawaiian Islands and in the Pacific pelagic sediments. Science **163**, 277 (1969).

REYNOLDS, J. H., VERHOOGEN, J.: Natural variations in the isotopic constitution of silicon. Geochim. Cosmochim. Acta **3**, 224 (1953).

RICKE, W.: Präparation von Schwefeldioxid zur massenspektrometrischen Bestimmung des S-Isotopenverhältnisses in natürlichen S-Verbindungen. Z. Anal. Chemie **199**, 401 (1964).

RIECK, G. R.: Einführung in die Massenspektroskopie. Berlin: VEB Deutscher Verlag der Wissenschaften 1956.

RITTENBERG, D.: The effect of isotopic substitution on the living cell. J. Chem. Phys. **60**, 318 (1963).

ROGINSKI, S. S.: Theoretische Grundlagen der Isotopenchemie. Berlin: VEB Deutscher Verlag der Wissenschaften 1962.

RONOV, A. B.: Organic carbon in sedimentary rocks (in relation to the presence of petroleum). Geochemistry 1959, 510—536.

RONOV, A. B.: Probable changes in the composition of seawater during the course of geologic time. Sedimentology 10, 25 (1968).

RONOV, A. B., MIGDISOV, A. A.: Geochemical history of the crystalline basement and the sedimentary cover of the Russian and North American platforms. Sedimentology 16, 137 (1971).

ROSENFELD, W., SILVERMAN, S.: Carbon isotope fractionation in bacterial production of methane. Science 130, 1658 (1959).

ROBERTSON, D. K., CUMMING, G. L.: Lead- and sulfur-isotope ratios from the Great Slave Lake area, Canada. Can. J. Earth Sci. 5, 1269 (1968).

RÖSLER, H. J., PILOT, J., HARZER, D., KRÜGER, P.: Isotopengeochemische Untersuchungen (O, S, C) an Salinar- und Sapropelsedimenten Mitteleuropas. 23. International Geological Congress. Vol. 6, 89 (1968).

RYE, R. O.: The carbon, hydrogen and oxygen isotopic composition of the hydrothermal fluids responsible for the lead-zinc deposits at Providencia, Zacatecas, Mexico. Econ. Geol. 61, 1399—1427 (1966).

RYE, R. O., O'NEIL, J. R.: The O^{18} content of water in primary fluid inclusions from Providencia, North Central Mexico. Econ. Geol. 63, 232—238 (1968).

RYE, R. O., CZAMANSKE, G. K.: Experimental determination of sphalerite-galena sulfur isotope fractionation and application to the ores at Providencia, Mexico. GSA meeting, program (1969).

RYZNAR, G., CAMPBELL, F. A., KROUSE, H. R.: Sulfur isotopes and the origin of the Quermont ore body. Econ. Geol. 62, 664—678 (1967).

SACKETT, W. M.: Carbon isotope composition of natural methane occurrences. Bull. Am. Ass. Petrol. Geol. 52, 853 (1968).

SACKETT, W. M., THOMPSON, R. R.: Isotopic organic carbon composition of recent continental derived clastic sediments of Eastern Gulf Coast, Gulf of Mexico. Bull. Am. Ass. Petrol. Geol. 47, 525 (1963).

SACKETT, W. M., ECKELMANN, W. R., BENDER, M. L., BE, A. W.: Temperature dependence of carbon isotope composition in marine plankton and sediments. Science 148, 235 (1965).

SACKETT, W. M., MOORE, W. S.: Isotopic variations of dissolved inorganic carbon. Chem. Geol. 1, 323—328 (1966).

SAKAI, H.: Fractionation of sulphur isotopes in nature. Geochim. Cosmochim. Acta 12, 150 (1957).

SAKAI, H.: Isotopic properties of sulfur compounds in hydrothermal processes. Geochemical J. 2, 29—49 (1968).

SAKAI, H., NAGASAWA, H.: Fractionation of sulfur isotopes in volcanic gases. Geochim. Cosmochim. Acta 15, 32 (1958).

SALOMONS, W.: Isotope fractionation between galena and pyrite and between pyrite and elemental sulfur. Earth Planet. Sci. Letters 11, 236 (1971).

SANGSTER, D. F.: Relative sulphur isotope abundances of ancient seas and stratabound sulphide deposits. Geol. Ass. Can. Proc. 19, 79—91 (1968).

SASAKI, A.: Sulphur isotope study of the Muskox intrusion, District of Mackenzie. Geol. Survey Canada, Paper 68—46 (1969).

SASAKI, A.: Seawater sulfate as a possible determinant for sulfur isotopic compositions of some stratabound sulfide ores. Geochem. J. 4, 41 (1970).

SASAKI, A., KROUSE, H. R.: Sulfur isotopes and the Pine Point lead-zinc mineralization. Econ. Geol. 64, 718—730 (1969).

SAVIN, S. M., EPSTEIN, S.: The oxygen and hydrogen isotope geochemistry of clay minerals. Geochim. Cosmochim. Acta **34**, 25—42 (1970a).

SAVIN, S. M., EPSTEIN, S.: The oxygen and hydrogen isotope geochemistry of ocean sediments and shales. Geochim. Cosmochim. Acta **34**, 43—64 (1970b).

SAVIN, S. M., EPSTEIN, S.: The oxygen isotopic compositions of coarse grained sedimentary rocks and minerals. Geochim. Cosmochim. Acta **34**, 323—329 (1970c).

SCALAN, R. C.: Ph. D. Thesis, University of Arkansas 1957.

SCHIEGL, W. E., VOGEL, J. C.: Deuterium content of organic matter. Earth Planet. Sci. Letters **7**, 307—313 (1970).

SCHILLER, W. R., GEHLEN, K. VON, NIELSEN, H.: Hydrothermal exchange and fractionation of sulfur isotopes in synthesized ZnS and PbS. Econ. Geol. **65**, 350 (1970).

SCHOEN, R., RYE, R. O.: Sulfur isotope distribution in solfataras, Yellowstone National Park. Science **170**, 1002 (1970).

SCHNEIDER, A.: The sulfur isotope composition of basaltic rocks. Contr. Mineral. Petrol. **24**, 95 (1970).

SCHNEIDER, A., NIELSEN, H.: Zur Genese des elementaren Schwefels im Gips von Weenzen. Contr. Mineral. Petrol. **11**, 705 (1965).

SCHREINER, G. D. L., VERBEEK, A. A.: Variations in $^{39}K/^{41}K$ ratio and movement of potassium in granite-shale contact region. Proc. Roy. Soc. London **A 285**, 423 (1965).

SCHREINER, G. D. L., WELKE, H. J.: Variations in $^{39}K/^{41}K$ ratio and movement of potassium in heated and stressed xenoliths. Geochim. Cosmochim. Acta **35**, 719 (1971).

SCHWARCZ, H. P.: Oxygen and carbon isotopic fractionation between coexisting calcite and dolomite. J. Geol. **74**, 38—48 (1966).

SCHWARCZ, H. P.: The stable isotopes of carbon. In: WEDEPOHL, K. H. (Ed.): Handbook of Geochemistry, 6-B-I. Berlin-Heidelberg-New York: Springer 1969.

SCHWARCZ, H. P., CLAYTON, R. N.: Oxygen isotope studies of amphibolites. Can. J. Earth Sci. **2**, 72—84 (1965).

SCHWARCZ, H. P., AGYEI, E. K., MCMULLEN, C. C.: Boron isotopic fractionation during clay adsorption from sea-water. Earth Planet. Sci. Letters **6**, 1—5 (1969).

SCHWARCZ, H. P., CLAYTON, R. N., MAYEDA, T.: Oxygen isotopic studies of calcareous and pelitic metamorphic rocks, New England. Bull. Geol. Soc. Am. **81**, 2299 (1970).

SENFTLE, F. E., BRACKEN, J. T.: Theoretical effect of diffusion on isotopic abundance ratios in rocks and associated fluids. Geochim. Cosmochim. Acta **7**, 61 (1955).

SHACKLETON, N.: Oxygen isotope analyses and Pleistocene temperatures re-assessed. Nature **215**, 15 (1967).

SHACKLETON, N.: Depth of pelagic foraminifera and isotopic changes in Pleistocene Oceans. Nature **218**, 79 (1968).

SHACKLETON, N.: The last interglacial in marine and terrestrial records. Proc. Roy. Soc. London **B 174**, 135 (1969).

SHARMA, T., CLAYTON, R. N.: Measurement of O^{18}/O^{16} ratios of total oxygen of carbonates. Geochim. Cosmochim. Acta **29**, 1347—1354 (1965).

SHEPPARD, S. M. F., EPSTEIN, S.: D/H and O^{18}/O^{16} ratios of minerals of possible mantle or lower crustal origin. Earth Planet. Sci. Letters **9**, 232 (1970).

SHEPPARD, S. M. F., NIELSEN, R. L., TAYLOR, H. P.: Oxygen and hydrogen isotope ratios of clay minerals from Porphyry Copper deposits. Econ. Geol. **64**, 755—777 (1969).

SHEPPARD, S. M. F., SCHWARCZ, H. P.: Fractionation of carbon and oxygen isotopes and magnesium between coexisting metamorphic calcite and dolomite. Contr. Mineral. Petrol. **26**, 161—198 (1970).

SHEPPARD, S. M. F., NIELSEN, R. L., TAYLOR, H. P.: Hydrogen and oxygen isotope ratios in minerals from Porphyry Copper Deposits. Econ. Geol. **66**, 515 (1971).

SHERGINA, Y. P., SHKORBATOV, S. S., KAMINSKAYA, A. B.: Relation of boron isotope composition to the magmatic source for basic and ultrabasic intrusions. Geochemistry **5**, 936 (1968).

SHIEH, Y. N., TAYLOR, H. P.: Oxygen and hydrogen isotope studies of contact metamorphism. Contr. Mineral. Petrol. **20**, 306—356 (1969 a).

SHIEH, Y. N., TAYLOR, H. P.: Oxygen and carbon isotope studies of contact metamorphism of carbonate rocks. J. Petrol. **10**, 307—331 (1969 b).

SHIEH, Y. N., SCHWARCZ, H. P.: Oxygen isotope studies on some granitic rocks from the Grenville Province of Southeastern Ontario. Trans. Am. Geophys. Union **52**, 365 (1971).

SHIMA, M.: Geochemical study of boron isotopes. Geochim. Cosmochim. Acta **27**, 911 (1963).

SHIMA, M.: The isotopic composition of magnesium in terrestrial samples. Bull. Chem. Soc. Japan **37**, 284 (1964).

SHIMA, M., HONDA, M.: Isotopic abundance of meteoritic lithium. J. Geophys. Res. **68**, 2849 (1963).

SHIMA, M., HONDA, H.: Distribution and isotopic composition of lithium in stone meteorites. Geochemical J. **1**, 27 (1966).

SHIMA, M., GROSS, W. H., THODE, H. G.: Sulfur isotope abundances in basic sills, differentiated granites and meteorites. J. Geophys. Res. **68**, 2835—2846 (1963).

SILVERMAN, S. R.: The isotope geology of oxygen. Geochim. Cosmochim. Acta **2**, 26—42 (1951).

SILVERMAN, S. R.: Investigations of petroleum origin and evolution mechanisms by carbon isotope studies. In: Isotopic and cosmic chemistry, p. 92. Amsterdam: North Holland Publ. Co. 1964.

SILVERMAN, S. R.: Carbon isotopic evidence for the role of lipids in petroleum. J. Am. Oil Chem. Soc. **44**, 691 (1967).

SILVERMAN, S. R., EPSTEIN, S.: Carbon isotopic compositions of petroleums and other sedimentary organic materials. Bull. Am. Ass. Petrol. Geol. **42**, 992 (1958).

SMITH, B. N., EPSTEIN, S.: Biogeochemistry of the stable isotopes of hydrogen and carbon in salt marsh biota. Plant. Physiol. **46**, 738 (1970).

SMITH, B. N., EPSTEIN, S.: Two categories of $^{13}C/^{12}C$ ratios for higher plants. Plant. Physiol. **47**, 380 (1971).

SMITH, J. W., KAPLAN, I. R.: Endogenous carbon in carbonaceous meteorites. Science **167**, 1367 (1970).

SMITHERINGALE, W. G., JENSEN, M. L.: Sulfur isotopic composition of the Triassic igneous rocks of Eastern United States. Geochim. Cosmochim. Acta **27**, 1183—1207 (1963).

SPAETH, C., HOEFS, J., VETTER, U.: The isotopic composition of belemnites and related paleotemperatures. Bull. Geol. Soc. Am. **82**, 3139 (1971).

SPEELMAN, E. L., SCHWARCZ, H. P.: Metamorphic sulfur isotope studies in the Haliberton-Madoc Area, Grenville Subprovince, Canada. Geol. Soc. Am. Ann. Meeting **1966**, 209.

SPOTTS, J. H., SILVERMAN, S. R.: Organic dolomite from Point Fermin, California. Amer. Min. **51**, 1144—1155 (1966).

STAHL, W.: Search for natural variations in calcium isotope abundances. Earth Planet. Sci. Letters **5**, 171 (1968).

STAHL, W., WENDT, I.: Fractionation of calcium isotopes in carbonate precipitation. Earth Planet. Sci. Letters **5**, 184 (1968).

STANTON, R. L.: The application of sulphur isotope studies in ore genesis theory. A suggested model. N.Z.J. Geol. Geophys. **3**, 375—389 (1960).

STANTON, R. L., RAFTER, T. A.: The isotopic constitution of sulphur in some stratiform lead-zinc ores. Mineral. Deposita **1**, 16—29 (1966).

STEVENS, L. M., KROUT, L., WALLING, D., VENTERS, A., ENGELKEMEIER, A., ROSS, L. E.: The isotopic composition of atmospheric carbon monoxide. Earth Planet. Sci. Letters **16**, 147—165 (1972).

SUZUOKI, T., EPSTEIN, S.: Hydrogen isotope fractionation factor (δ's) between muscovite, biotite, hornblende and water (Abstr.). Eos (Trans. Am. Geophys. Union) **51**, 451 (1970).

TAKEMATSU, N., MATSUO, S., SATO, S.: Isotopic composition of magnesium in upper mantle material and a meteorite. Geochemical J. **1**, 51 (1967).

TAN, F. C., HUDSON, J. D.: Carbon and oxygen isotopic relationships of dolomites and coexisting calcites, Great Estuarine Series (Jurassic), Scotland. Geochim. Cosmochim. Acta **35**, 755 (1971).

TARUTANI, T., CLAYTON, R. N., MAYEDA, T. K.: The effect of polymorphism and magnesium substitution on oxygen isotope fractionation between calcium carbonate and water. Geochim. Cosmochim. Acta **33**, 987 (1969).

TATSUMI, T.: Sulfur isotopic fractionation between coexisting sulfide minerals from some Japanese ore deposits. Econ. Geol. **60**, 1645—1659 (1965).

TAYLOR, H. P.: Isotopic evidence for large-scale oxygen exchange during metamorphism of Adirondack igneous rocks (abstract). Geol. Soc. Am. Spec. Paper **76**, 163 (1964).

TAYLOR, H. P.: Oxygen isotope studies of hydrothermal mineral deposits. In: BARNES, H. L. (Ed.): Geochemistry of hydrothermal ore deposits. New York-London: Holt, Rinehart and Winston Inc. 1967.

TAYLOR, H. P.: The oxygen isotope geochemistry of igneous rocks. Contr. Mineral. Petrol. **19**, 1—71 (1968a).

TAYLOR, H. P.: Oxygen isotope studies of anorthosites with special reference to the origin of bodies in the Adirondack Mountains, New York. In: Origin of anorthosites, N.Y. State Museum Spec. Publ. 1968b.

TAYLOR, H. P.: Oxygen isotope evidence for large-scale interaction between meteoric ground waters and Tertiary granodiorite intrusions, Western Cascade Range, Oregon. J. Geophys. Res. **76**, 7855 (1971).

TAYLOR, H. P., EPSTEIN, S.: $^{18}O/^{16}O$ ratios of feldspars and quartz in zoned granitic pegmatites. Geol. Soc. Am. Spec. Paper **68**, 283 (1961).

TAYLOR, H. P., EPSTEIN, S.: Relation between O^{18}/O^{16} ratios in coexisting minerals of igneous and metamorphic rocks. I. Principles and experimental results. Bull. Geol. Soc. Am. **73**, 461 (1962a).

TAYLOR, H. P., EPSTEIN, S.: Relation between O^{18}/O^{16} ratios in coexisting minerals of igneous and metamorphic rocks. II. Application to petrologic problems. Bull. Geol. Soc. Am. **73**, 675 (1962b).

TAYLOR, H. P., EPSTEIN, S.: O^{18}/O^{16} ratios in rocks and coexisting minerals of the Skaergaard intrusion. J. Petrol. **4**, 51 (1963).

TAYLOR, H. P., EPSTEIN, S.: Comparison of oxygen isotope analyses of tektites, soils and impactite glasses. In: Cosmic and Isotopic Chemistry. Amsterdam: North Holland Pull. Co. 1964.

TAYLOR, H. P., EPSTEIN, S.: Oxygen isotope studies of Ivory Coast tektites and impactite glass from the Bosumtwi crater, Ghana. Science **153**, 173 (1966a).

TAYLOR, H. P., EPSTEIN, S.: Deuterium-hydrogen ratios in coexisting minerals of metamorphic and igneous rocks (abstract). Trans. Am. Geophys. Union **47**, 213 (1966b).

TAYLOR, H. P., EPSTEIN, S.: Hydrogen isotope evidence for influx of meteoric ground water into shallow igenous intrusives (abstract). Geol. Soc. Am. Annual Meeting, Mexico City **1968**, 294.

TAYLOR, H. P., EPSTEIN, S.: Correlations between O^{18}/O^{16} ratios and chemical compositions of tektites. J. Geophys. Res. **74**, 6834—6844 (1969).

TAYLOR, H. P., ALBEE, A. L., EPSTEIN, S.: O^{18}/O^{16} ratios of coexisting minerals in three assemblages of kyanite-zone pelitic schists. J. Geol. **71**, 513 (1963).

TAYLOR, H. P., DUKE, M. B., SILVER, L. T., EPSTEIN, S.: Oxygen isotope studies of minerals in stony meteorites. Geochim. Cosmochim. Acta **29**, 489 (1965).

TAYLOR, H. P., FRECHEN, J., DEGENS, E. T.: Oxygen and carbon isotope studies of carbonatites from the Laacher See district, West Germany and the Alno district, Sweden. Geochim. Cosmochim. Acta **31**, 407 (1967).

TAYLOR, H. P., COLEMAN, R. G.: O^{18}/O^{16} ratios of coexisting minerals in glaucophane-bearing metamorphic rocks. Geol. Soc. Am. Bull. **79**, 1727—1756 (1968).

TAYLOR, H. P., FORESTER, R. W.: Low-^{18}O igneous rocks from the intrusive complexes of Skye, Mull and Ardnamurchan, Western Scotland. J. Petrol. **12**, 465 (1971).

THODE, H. G.: Sulphur isotope geochemistry. In: SHAW, D. M. (Ed.): Studies in analytical geochemistry, p. 25. Toronto: University of Toronto Press 1963.

THODE, H. G.: Sulphur isotope geochemistry fractionation between coexisting sulphide minerals. Geol. Soc. Am. Mem. (in press) (1972).

THODE, H. G., MAĆNAMARA, J., COLLINS, C. B.: Natural variations in the isotopic content of sulphur and their significance. Can. J. Res. **27 B**, 361 (1949).

THODE, H. G., KLEEREKOPER, H., McELCHERAN, D.: Isotope fractionation in the bacterial reduction of sulphate. Research **4**, 581 (1951).

THODE, H. G., WANLESS, R. K., WALLOUCH, R.: The origin of native sulphur deposits from isotope fractionation studies. Geochim. Cosmochim. Acta **5**, 286—298 (1954).

THODE, H. G., MONSTER, J., DUNFORD, H. B.: Sulphur isotope abundances in petroleum and associated materials. Am. Ass. Petrol. Geol. Bull. **42**, 2619 (1958).

THODE, H. G., HARRISON, A. G., MONSTER, J.: Sulphur isotope fractionation in early diagenesis of recent sediments of Northeast Venezuela. Bull. Am. Ass. Petrol. Geol. **44**, 1809 (1960).

THODE, H. G., MONSTER, J., DUNFORD, H. B.: Sulphur isotope geochemistry. Geochim. Cosmochim. Acta **25**, 159—174 (1961).

THODE, H. G., DUNFORD, H. B., SHIMA, M.: Sulphur isotope abundances in rocks of the Sudbury District and their geological significance. Econ. Geol. **57**, 565—578 (1962).

THODE, H. G., MONSTER, J.: The sulfur isotope abundances in evaporites and in ancient oceans. In: VINOGRADOV, A. P. (Ed.): Proc. Geochemical Conf. Commemorating the Centenary of V.I. Vernadskii's Birth. **2**, 630 (1964).

THODE, H. G., MONSTER, J.: Sulphur isotope geochemistry of petroleum, evaporites and ancient seas. In: Fluids in subsurface environments — a symposium: Am. Ass. Petrol. Geol. Mem. **4**, 367 (1965).

THODE, H. G., MONSTER, J.: Sulfur isotope abundances and genetic relations of oil accumulations in Middle East Basins. Am. Ass. Petrol. Geol. Bull. **54**, 627 (1970).

THODE, H. G., REES, C. E.: Die Geochemie der Schwefelisotope bei den Erdölvorkommen des mittleren Ostens. Endeavour **29**, 24—28 (1970).

THODE, H. G., CRAGG, C. B., HULSTON, J. R., REES, C. E.: Sulphur isotope exchange between sulphur dioxide and hydrogen sulphide. Geochim. Cosmochim. Acta **35**, 35 (1971).

TILLES, D.: Natural variations in isotopic abundances of silicon. J. Geophys. Res. **66**, 3003 (1961 a).

TILLES, D.: Variations of silicon isotope ratios in a zoned pegmatite. J. Geophys. Res. **66**, 3015 (1961 b).

TILLES, D.: Stable silicon isotope ratios in tektites. Geochim. Cosmochim. Acta **28**, 1015 (1964).

TOURTELOT, H. A., RYE, R. O.: Distribution of oxygen and carbon isotopes in fossils of late Cretaceous age, Western Interior Region of North America. Bull. Geol. Soc. Am. **80**, 1903—1922 (1969).

TRASK, P. O., PATNODE, H. W.: Recent marine sediments, a symposium, ed. by P.O. Trask. Oklahoma: Tulsa 1939.

TROFIMOV, A.: Isotopic constitution of sulfur in meteorites and in terrestrial objects (in Russian). Dokl. Akad. Nauk SSSR **66**, 181 (1949).

TUDGE, A. P.: A method of analysis of oxygen isotopes in orthophosphate — its use in the measurement of paleotemperatures. Geochim. Cosmochim. Acta **18**, 81—93 (1960).

TUDGE, A. P., THODE, H. G.: Thermodynamic properties of isotopic compounds of sulphur. Can. J. Res. **28**, 567—579 (1950).

TUPPER, W. M.: Sulfur isotopes and the origin of sulfide deposits of the Bathurst Newcastle area of Northern New Brunswick. Econ. Geol. **55**, 1676—1707 (1960).

TURI, B., TAYLOR, H. P.: An oxygen and hydrogen isotope study of a granodiorite pluton from the Southern California batholith. Geochim. Cosmochim. Acta **35**, 383—406 (1971 a).

TURI, B., TAYLOR, H. P.: $^{18}O/^{16}O$ ratios of the Johnny Lyon Granodiorite and Texas Canyon Quartz Monzonite Plutons, Arizona and their contact aureoles. Contr. Mineral. Petrol. **32**, 138 (1971 b).

UPHAUS, R. A., KATZ, J. J.: Deuterium isotope effect on carbon isotope fractionation in photosynthesis. Science **155**, 324 (1967).

UREY, H. C.: The thermodynamic properties of isotopic substances. J. Chem. Soc. **1947**, 562.

UREY, H. C., BRICKWEDDE, F. G., MURPHY, G. M.: An isotope of hydrogen of mass 2 and its concentration (abstract). Phys. Rev. **39**, 864 (1932 a).

UREY, H. C., BRICKWEDDE, F. G., MURPHY, G. M.: A hydrogen isotope of mass 2 and its concentration. Phys. Rev. **40**, 1 (1932 b).

UREY, H. C., LOWENSTAM, H. A., EPSTEIN, S., McKINNEY, C. R.: Measurement of paleotemperatures and temperatures of the Upper Cretaceous of England, Denmark and the Southeastern United States. Bull. Geol. Soc. Am. **62**, 399 (1951).

VERHOOGEN, F. J., WEISS, L. E., WAHRHAFTIG, C., FYFE, W. S.: The Earth. New York-London: Holt, Rinehart and Winston 1970.

VERBEEK, A. A., SCHREINER, G. D.: Variations in $^{39}K/^{41}K$ ratios and movement of potassium in a granite-amphibolite contact region. Geochim. Cosmochim. Acta **31**, 2125—2133 (1967).

VINOGRADOV, A. P., CHUPAKHIN, M. S., GRINENKO, V. A.: The S^{32}/S^{34} isotope ratios in sulfides. Geochemistry **1956**, No. 4, 3.

VINOGRADOV, A. P., CHUPAKHIN, M. S., GRINENKO, V. A.: Some data on the isotopic composition of sulfur in sulfides. Geochemistry **1957**, 183.

VINOGRADOV, A. P., DONTSOVA, E. I., CHUPAKHIN, M. S.: Isotopic ratios of oxygen in meteorites and igneous rocks. Geochim. Cosmochim. Acta **18**, 278—293 (1960).

VINOGRADOV, A. P., GRINENKO, V. A., USTINOV, V. I.: Isotopic composition of sulfur compounds in the Black Sea. Geochemistry **1962**, 973—997.

VINOGRADOV, A. P., KROPOTOVA, O. I., USTINOV, V. I.: Possible sources of carbon in diamonds as indicated by the C^{12}/C^{13} ratios. Geochemistry **1965**, 495—503.

VINOGRADOV, A. P., KROPOTOVA, O. I., ORLOV, Y. L., GRINENKO, V. A.: Isotopic composition of diamond and carbonado crystals. Geochemistry **1966**, 1123.

VINOGRADOV, A. P., KROPOTOVA, O. I., VDOVYKIN, G. D., GRINENKO, V. A.: Isotopic composition of different phases of carbon in carbonaceous meteorites. Geochemistry **1967**, 229—235.

VOGEL, J. C.: Über den Isotopengehalt des Kohlenstoffs in Süsswasser-Kalkablagerungen. Geochim. Cosmochim. Acta **16**, 236 (1959).

VOGEL, J. C.: Isotope separation factors of carbon in the equilibrium system CO_2-HCO_3-CO_3. In: Summer course on Nuclear Geology Varenna 1960, published Pisa 1961.

VOGEL, D. E., GARLICK, G. D.: Oxygen-isotope ratios in metamorphic eclogites. Contr. Mineral. Petrol. **28**, 183—191 (1970).

WALTER, L. S., CLAYTON, R. N.: Oxygen isotopes: experimental vapor fractionation and variations in tektites. Science **156**, 1357 (1967).

WANLESS, R. K., BOYLE, R. W., LOWDON, J. A.: Sulfur isotope investigations on the gold-quartz deposits of the Yellowknife District. Econ. Geol. **55**, 1591—1621 (1960).

WASSERBURG, G. J., MAZOR, E., ZARTMAN, R. H.: Isotopic and chemical composition of some terrestrial natural gases. In: GEISS, T., GOLDBERG, E. D. (Eds.): Earth science and meteoritics. Amsterdam: North-Holland Publ. Co. 1963.

WAY, K., FANO, L., SCOTT, M. R., THEW, K.: Nuclear data. A collection of experimental values of halflifes, radiation energies, relative isotopic abundances, nuclear moments and cross-sections. Natl. Bur. Standards U.S. Circ. 499 (1950).

WEBER, J. N.: Oxygen isotope fractionation between coexisting calcite and dolomite. Science **145**, 1303 (1964).

WEBER, J. N.: Oxygen isotope fractionation between coexisting calcite and dolomite in the freshwater Upper Carboniferous Freeport Formation. Nature **207**, 972 (1965).

WEBER, J. N.: Possible changes in the isotopic composition of the oceanic and atmospheric carbon reservoir over geologic time. Geochim. Cosmochim. Acta **31**, 2343 (1967).

WEBER, J. N.: Fractionation of the stable isotopes of carbon and oxygen in calcareous marine invertebrates — the Asteroidea, Ophiuroidea and Crinoidea. Geochim. Cosmochim. Acta **32**, 33—70 (1968).

WEBER, J. N., WILLIAMS, E. G., KEITH, M. L.: Paleoenvironmental significance of carbon isotopic composition of siderite nodules in some shales of Pennsylvanian age. J. Sediment. Petrol. **34**, 814 (1964).

WEBER, J. N., RAUP, D. M.: Fractionation of the stable isotopes of carbon and oxygen in marine calcareous organisms — the Echinoidea. I. Variation of ^{13}C and ^{18}O content within individuals. Geochim. Cosmochim. Acta **30**, 681—704 (1966a).

WEBER, J. N., RAUP, D. M.: Fractionation of the stable isotopes of carbon and oxygen in marine calcareous organisms — the Echinoidea. II. Environmental and genetic factors. Geochim. Cosmochim. Acta **30**, 705 (1966b).

WEDEPOHL, K. H.: Die Zusammensetzung der Erdkruste. Fortschr. Mineralog. **46**, 145 (1969).

WELLMANN, R. P., COOK, F. D., KROUSE, H. R.: Nitrogen-15: Microbiological alteration of abundance. Science **161**, 269 (1968).

WELTE, D. H.: Organischer Kohlenstoff und die Entwicklung der Photosynthese auf der Erde. Naturwissenschaften **57**, 17—23 (1970).

WENDT, I.: Fractionation of carbon isotopes and its temperature dependence in the system $CO_{2(gas)}$-CO_2 (solution) and HCO_3^--CO_2 (solution). Earth Planet. Sci. Letters **4**, 64—68 (1968).

WENNER, D. B., TAYLOR, H. P.: Temperatures of serpentinization of ultramafic rocks based on $^{18}O/^{16}O$ fractionation between coexisting serpentine and magnetite. Contr. Mineral. Petrol. **32**, 165 (1971).

WHITE, D. E.: Thermal waters of volcanic origin. Bull. Geol. Soc. Am. **68**, 1637 (1957a).

WHITE, D. E.: Magmatic, connate and metamorphic waters. Bull. Geol. Soc. Am. **68**, 1659 (1957b).

WHITE, D. E.: Environments of generation of some basemetal ore deposits. Econ. Geol. **63**, 301—335 (1968).

WHITE, F. A., COLLINS, T. L., ROURKE, F. M.: Search for possible naturally occurring isotopes of low abundance. Phys. Rev. **101**, 1786 (1956).

WICKMAN, F. E.: Variations in the relative abundance of the carbon isotopes in plants. Geochim. Cosmochim. Acta **2**, 243 (1952).

WICKMAN, F. E.: Wird das Häufigkeitsverhältnis der Kohlenstoffisotopen bei der Inkohlung verändert. Geochim. Cosmochim. Acta **3**, 249 (1953).

WICKMAN, F. E.: The cycle of carbon and the stable isotopes of carbon. Geochim. Cosmochim. Acta **9**, 136 (1956).

WILSON, A. F., GREEN, D. C., DAVIDSON, R. L.: The use of oxygen isotope geothermometry on the granulites and related intrusives, Musgrave Ranges, Central Australia. Contr. Mineral. Petrol. **27**, 166—178 (1970).

WLOTZKA, F.: Untersuchungen zur Geochemie des Stickstoffs. Geochim. Cosmochim. Acta **24**, 106 (1961).

WOODCOCK, A. H., FRIEDMAN, I.: The deuterium content of raindrops. J. Geophys. Res. **68**, 4477 (1963).

ZARTMAN, R. E., WASSERBURG, G. J., REYNOLDS, J. J.: Helium, argon and carbon in some natural gases. J. Geophys. Res. **66**, 277 (1961).

ZBOROWSKI, G., PONTICORVO, L., RITTENBERG, D.: On the constancy of deuterium fractionation in the biosynthesis of fatty acids since the Miocene period. Proc. Nat. Acad. Sci. U.S. **58**, 1660 (1967).

Subject Index

Minerals, Rocks and Inorganic Materials
